CAMBRIDGE COUNTY GEOGRAPHIES

SCOTLAND

General Editor: W. Murison, M.A.

FIFESHIRE

Cambridge County Geographies

FIFESHIRE

by

EASTON S. VALENTINE, M.A.

Headmaster of the English Department in Dundee High School:
Formerly Examiner in English in the University of Glasgow
and in the University of St Andrews

With Maps, Diagrams and Illustrations

Cambridge:
at the University Press
1910

CAMBRIDGE UNIVERSITY PRESS
Cambridge, New York, Melbourne, Madrid, Cape Town,
Singapore, São Paulo, Delhi, Mexico City

Cambridge University Press
The Edinburgh Building, Cambridge CB2 8RU, UK

Published in the United States of America by Cambridge University Press, New York

www.cambridge.org
Information on this title: www.cambridge.org/9781107662032

First published 1910
First paperback edition 2013

A catalogue record for this publication is available from the British Library

ISBN 978-1-107-66203-2 Paperback

CONTENTS

ILLUSTRATIONS

MAPS

The illustrations on pp. 5, 12, 16, 18, 20, 25, 32, 33, 41, 47, 49, 53, 55, 59, 62, 75, 79, 84, 91, 95, 101, 107, 110, 114, 115, 118, 119, 122, 123, 124, 125, 129, 130, 131, 133, 135, 137, 138, 139, 140, 142, 154, 160, 167, 169, 171, 174, 181, 182 are from photographs by Messrs Valentine and Sons, Limited, Dundee; and that on p. 179 is from a photograph by Messrs Wilson Brothers of Aberdeen.

For permission to reproduce the map on p. 146, the author is indebted to the courtesy of the Chairman of The Fife Tramway Light and Power Company, Limited; for leave to reproduce the picture on p. 117 to Messrs G. P. Putnam's Sons; the portrait on p. 158 is reproduced by kind permission of Prof. C. S. Terry and Messrs Longmans, Green & Co.

1. County and Shire. The Origin of Fife.

County and shire, names applied to various subdivisions of Great Britain and Ireland, are obviously of different derivation, and are not always interchangeable. *County* is a French word that originally signified a territory ruled over by, or in charge of, a count, and may be taken as the equivalent of the English *earldom* ; whereas *shire*, simply another form of *share*, would probably be first applied to the sections into which the old English kingdoms were divided when the whole land came to be ruled by a single monarch. *County* is pretty well applicable to any of the divisions of Great Britain and Ireland ; but *shire* is not used in Ireland, where we talk only of County Dublin, etc.

In the case of Kent and a few other very ancient divisions, the proper name alone, without either divisional designation, is usually employed; and this is the case with that portion of Scotland with which we are here concerned. We may speak, indeed, of either Fifeshire or the county of Fife, or, by ancient usage, even of the "kingdom of Fife"; but, like Kent, Fife is a name that often stands alone as sufficiently distinctive.

There is a reason for this. Fife seems to have been originally one of the kingdoms of the Picts; and, cut off as it is by water and mountain range from the rest of the country, its comparative isolation may have given it a kind of independence unknown to other parts of Pictland, although no central authority appears at any time to have existed in that region. "Fibh," pronounced and afterwards written Fife, is probably a Pictish name, with a meaning not very certainly known to us. Tradition has it that Fibh was one of the seven sons of Cruithne, an obscure chief or king. The name "Fif" first occurs in verses that have been ascribed to St Columba, where it is used as the name of one of the provinces of Pictavia.

Fife, which till the beginning of the fourteenth century included Kinross, was in the ninth century divided into two by a line running from Leven to Auchtermuchty—the eastern part called Fife, the western Fothrif. Still later it was divided into five parts, called quarters, Inverkeithing, Dunfermline, Leven, Eden and Crail. The ecclesiastical as well as the civil independence of Kinross was possibly established by Robert the Bruce. At all events, Kinross, in a smaller form since 1305 and in its modern form since 1685, has been regarded as a separate county.

Certain changes were made (1891) in Fife by the Boundary Commissioners. A part in the north of the parish of Dunfermline was transferred to Kinross-shire; while Fife acquired a detached portion of the parish of Portmoak. The Cut, or new channel of the river Leven, was ordained, for a certain part, to be the boundary between the two counties instead of the old bed of the river. More important still was the transference from Perthshire to Fifeshire of the parishes of Culross and Tulliallan, which had previously formed a portion of Perthshire separated therefrom by Clackmannanshire. Perth gained at the expense of Fife in the parish of Abernethy, most of which went to Perth, while the portion that remained in Fife was merged in the parish of Newburgh.

2. General Characteristics. Its Position and Natural Conditions.

Fife, the peninsula occupying a central position on the east coast of Scotland, is admittedly one of the most interesting counties in Britain. In few districts of equal area can be found so much to engross the attention of the historian, the antiquarian, the geologist, or the mining engineer, while its sea-bathing places are resorted to by thousands from far and near, and its golf-links are of world-wide celebrity.

In early days, while America was yet unknown, and before trade with that continent and the natural resources of Clydesdale had made Glasgow the commercial centre of Scotland, Fife and the Lothians were undoubtedly the parts of our country that attracted the chief attention of natives and foreigners alike. And Fife was essentially the royal county before Midlothian, Dunfermline the capital before Edinburgh. Long a favourite resort of our Scottish kings, who held court at Dunfermline and hunted on the slopes of the Lomonds and in the woods of Falkland, Fife is remarkable for the extraordinarily large number of its royal burghs. These still hold their ancient charters though they suffered in importance, as did the county in general, from the transference of the court, first to Edinburgh, and afterwards to London. Celtic influence and Culdee Christianity came from the north-west, Anglian influence and Roman Christianity from the south-east, and these influences and civilizations met and coalesced around the shores of the Forth. Standing near the site of Malcolm Canmore's Tower in Pittencrieff Park, Dunfermline, and looking south to the castle-crowned Rock of Edinburgh clearly visible in the distance beyond the blue waters of the great firth, one may take in at a single glance this classic ground of Scottish story. For centuries, too, ecclesiastical and educational interest were focused in St Andrews with its hoary cathedral and ancient university. "Fife," it has been said, "contains the concentrated essence of Scottish history and character."

The ancient rank of the county is still marked for the antiquary by an unusually large number of ruins of abbeys, churches and castles, while in few districts are there so many baronial residences and gentlemen's seats.

For its natural characteristics and vast mineral resources, Fifeshire is not less remarkable. "If I were asked," says Sir Archibald Geikie, "to select a region in the British Isles where geology could best be practically taught by constant appeals to evidence in the field, I

St Andrews, from South

would with little hesitation recommend the east of Fife as peculiarly adapted for such a purpose." Its striking shore-sections, its volcanic necks, its raised beaches, its fossils, kames, erratic blocks, and vanished lakes make Fife a unique geological region.

Again, the mineral resources of the county are both varied and abundant. Its coalfield was the first in Scotland to be worked; and the modern machinery for boring and pumping, which renders the attainment of greater depths in mining practicable, is rapidly making it the premier coal district in North Britain. This industry has been greatly fostered by the opening of important docks at Methil, Burntisland, and other Fifeshire ports. Building stone of excellent quality is quarried in various parts, and there is an inexhaustible supply of material for constructing docks and piers and metalling roads. Other minerals found within its boundaries are also capable of economic development.

Various important manufactures are carried on in Fifeshire, such as paper-making and distilling, and on a very large scale, at Kirkcaldy and Dunfermline respectively, are produced floorcloths and fine linens that find ready markets all over the world.

The fishing industry, which has its headquarters at Anstruther, is distributed over about a score of small towns and villages dotted along a coast-line nearly 115 miles in length, from Kincardine-on-Forth to Newburgh-on-Tay, while the produce of net and line and creel finds its way to Germany, Russia, and even Central Asia.

The farming of Fife is both varied and successful. Its

cereals are amongst the best in Scotland, for the farmers of "The Kingdom" have put to good use the naturally fertile portions, and have adopted the best methods and the newest agricultural appliances so as to render less promising tracts productive. The rearing of horses, cattle, and sheep is also extensively carried on.

Although Fifeshire is cut off by the Ochils from the valley of Strathmore—the natural avenue of commerce between central and north-eastern Scotland—the enterprise of the North British Railway Company has not only carried a railroad through Glenfarg, but by throwing across the broad estuaries of the Forth and Tay two of the longest and greatest bridges of the world, and thus extending its system from Berwick to Aberdeen, it has freed Fife from its isolation and linked it, north and south, in a continuous route that is the most direct between the east of Scotland and London. The Forth Bridge is indeed one of the greatest marvels of modern engineering.

The need of coast defence has called for recent attention on the part of the Government. The Tay, with Dundee as its headquarters, has become the station of a submarine flotilla, and at Rosyth, just above the Bridge on the Fifeshire side of the Forth, the construction of a naval base that promises to be a Scottish Portsmouth, is being rapidly carried on. All this will undoubtedly tend to increase the importance of a county that is already the scene of an unusual variety of activities.

3. Size. Shape. Boundaries.

Fife is bounded on the west by the counties of Clackmannan, Kinross, and Perth, being separated from the last of these by the Ochil Hills. It is washed on the north, east, and south respectively by the waters of the Firth of Tay, the German Ocean, and the Firth of Forth. The trend of the land is north-easterly. Before Kinross was made into a separate county, the whole shire formed a rough parallelogram, whose two seaward angles are the promontories of Tentsmuir Point and Fife Ness. The regularity of its outline has, however, been deeply notched in the west by the artificial excision of Kinross-shire, and on the east by the natural indentation of St Andrews Bay with the small estuary of the river Eden. The south shore of the Firth of Tay presents an almost unbroken straight line from Newburgh to Ferry-Port-on-Craig, between which and Tentsmuir Point there is a small bay of shallow water. The comparative uniformity of the coast-line between Fife Ness and North Queensferry, where the county reaches its southernmost point, is broken by a sweeping curve between Pettycur and Elie that forms towards the east the bow-shaped Bay of Largo. On this part of the coast there stand the chief ports of the county. Between Pettycur and Queensferry the shore is more deeply indented by a succession of small bays. At North Queensferry occurs the geologically remarkable peninsula that rendered practicable the bridging of the Forth. Beyond the Bridge to Kincardine, in Tulliallan, the most westerly of the Fife parishes, the coast is fairly regular, except for the small curve of Torry Bay. The western boundary is extremely irregular, following now a line of hills like those of Cleish and Benarty, now small local streams that flow west into the river Devon or descend from the declivities of the Ochils into Loch Leven. In many places it is artificial.

The length of the county from Fife Ness to the extreme south-west is 41½ miles; its breadth from Newburgh to Burntisland 21 miles; its coast-line on estuary and sea, if the curvature is taken into account, is about 115 miles; and the length of its landward boundary, ins and outs considered, is 61 miles. This long line of 176 miles encloses an area of 328,427 acres, or 513 square miles.

In point of size Fife is the sixteenth Scottish county, and its area is about one-sixtieth of the whole country. While thus far out-distanced in size by many shires, only a few, notably Aberdeen, Perth, and Lanark, of those whose area is larger, exceed it in industrial importance; while of the smaller counties, only Stirling, Edinburgh and Renfrew can vie with it as scenes of historical interest or commercial activity.

4. Surface and General Features.

The first thing to learn about Fife, the key to all the rest, is its land elevation. While in some cases lofty enough to be designated mountains in a strictly technical sense—for a good many points are more than 1000 feet above sea-level—the heights of Fife, in contrast to the higher Grampians, are usually spoken of as hills, and the county forms part of the Lowlands of Scotland. Its length is traversed by two ranges of a fairly continuous character, which are generally highest in the west and gradually diminish as they approach the sea. In the west and north are the Ochils, which enter Fife from Perth but have left their loftiest summits (2363 feet) far behind them before they reach our county. The hills that form the high ground of the north extend from Newburgh to Tayport with a breadth in the west of some six miles and slope steeply northwards, leaving a narrow foreshore along the margin of the river. Many of their heights are wooded, others grassy, and often they terminate in rocky crests. Among the chief summits are Lumbenny Hill (889 feet) behind Newburgh; Norman's Law (850 feet) farther to the east, a striking hill that commands a magnificent view of the valley of the Tay and the mountains beyond; and Lucklaw Hill (626 feet) near Leuchars. There are many lovely nooks within this region, and its well-made roads make walking or cycling a great pleasure. Occasionally one catches glimpses of the firth, here so land-locked as to have the appearance at places of a large lake; or the vista up some stream is closed by a crag crowned with trees.

Through the middle of the county runs a line of hills, which, though not perfectly continuous, forms its backbone. These are highest in the west and attain in some points even mountainous elevations. In the south-western reaches are Saline Hill (1178 feet), Knock Hill (1189 feet), and numerous lower heights; and to the north of these the Cleish Hills (1241 feet) and Benarty (1131 feet) separate Fife and Kinross. The latter lies directly south of Loch Leven, while to the east of that fine sheet of water rises the West Lomond (1731 feet), the highest summit in the county. It is visible from far and near, and commands an unrivalled view. Away to the north and the west tower the Highland peaks of Stirling, Dumbarton and Perth; north-east is the Firth of Tay with Dundee and the Sidlaws; to the south-east lies Edinburgh with its picturesque local heights and the background of the Pentland, the Moorfoot, and the Lammermuir Hills; while in the east sparkle the waters of the German Ocean. Situated nearly in the centre of the peninsula, the hill seems to dominate the whole county. It is connected by a lofty ridge with the East Lomond (1471 feet), an almost equally good point of view, from which it descends in a long eastward slope that is crossed between Markinch and Kettle by the main line of railway.

To the east of this the hills spread out in an irregular and somewhat bewildering manner, forming a tumbled district of heights and hollows. Clatto Hill (547 feet), Kellie Law (500 feet), and Carnbee Law continue the line of the Lomonds. On the south-eastern side of the valley of the Eden are Walton Hill (621 feet), Wemyss Hall Hill (693 feet), Cairngreen, and Blebo; and farther east, in the direction of St Andrews, are Ladeddie Hill (600 feet), and Kinninmonth Hill (547 feet); besides which there are long ridges with a north-easterly trend, and an elevation of from 500 to 600 feet. The crests and upper slopes of these hills display rocky crags and steep escarpments. To the south, overlooking Largo Bay, and balancing, as it were, North Berwick Law in Haddington, towers the cone of Largo Law to a height of 965 feet.

East Lomond, Falkland

The marginal lands of the Forth are in some parts much broader than those of the Tay, but they are not by any means flat. Indeed there is a persistent tract of high ground behind the coast between Kirkcaldy and Inverkeithing, which forms a southern boundary to the valley of the Ore, and presents a picturesque line of heights along the shore. In Dunearn Hill an elevation of 727 feet is attained; and the remarkable hill known as the Binn of Burntisland is one of the most interesting in the whole county.

Throughout the county the trend of the ice during the glacial period, which was generally from west to east, has given a marked character to much of the higher ground. Escarpments and precipices face the west, and the ground behind falls away in a gradual slope eastwards, or merges into a smooth ridge following the same direction. Familiar instances in adjacent counties are the rocks on which are built the castles of Stirling and Edinburgh. Such formations are known as "crag and tail." The Isle of May, though now five miles from the coast, furnishes another example, and being at one time connected with the land must have had a similar origin.

The range of the Lomond Hills divides the county into two portions. Between the Ochils and their eastern continuation that flanks on the south the estuary of the Tay on the one hand, and the central watershed of the county on the other, lies the Howe of Fife, or valley of the Eden river, nearly 30 miles long. It varies in breadth. Near Cupar it is little more than a mile and a half across, but above that town it widens out in an expansion that may mark the site of a former lake. Between Auchtermuchty and Freuchie there is a stretch of nearly five miles, but so little does the plain rise that even ten miles farther west it is less than 200 feet above sea-level. Beyond Falkland the Lomonds send down a spur towards Strathmiglo that narrows the valley between that village and the Ochils to somewhat less than a mile.

To the south of the Lomonds and the west of the hills of the East Neuk lies the other great plain of Fife. This may roughly be regarded as the valley of the Leven and its main affluent, the Ore; while beyond Dunfermline the expanse is continued, though here a more hilly tract, in the district that slopes from the Saline, Cleish and neighbouring hills to the shore of the Forth.

A somewhat remarkable surface feature of Fife, especially in its eastern portion, is its numerous dens, or deans. These have been cut out of the solid rock by the age-long action of what may now appear quite inconsiderable streamlets, in which the declivity of the channel has compensated for volume of water. The dens are often from 70 to 100 feet in depth, while the cliffs on either side are not more than 300 feet apart. Fully a score might be enumerated, but we select only a few. Gowl's Den, near Kilmany, in the north of the county, is a deep little dell cut by a tiny brook in the lavas of the hills. Between West and East Newport the local railway bridges a romantic den through which a brook descends abruptly to the Tay at Newport Pier. Kittock's Den, near Boarhills, is a ravine 100 feet in depth, through which flows a rivulet with a course of only three-quarters of a mile; but in pre-glacial times this may have been the outlet of the larger Kenly Burn near by. Higher up that burn is a series of gorges that extend along its course for three miles. In the southern part of the Neuk on burns of corresponding names there are Kiel Den and

Hatton Den ; but the most important in the district are those of Balcarres and Balneil, ravines cut by the Kilconquhar Burn. On the Eden itself there are none of these erosion gorges, but its southern tributaries pass through several that are well known. None are, however, more celebrated than Dura Den on the Ceres Burn.

The "links" of Fife are famous, not because such ground is peculiar to this region, or because these tracts of land are more extensive here than elsewhere, but through being specially associated in Fifeshire with the game of golf. They occur in river valleys, but especially on low ground near the sea, where their soil is composed of sand or gravel which becomes gradually covered by a thin layer of vegetable matter on which grass grows freely. The best known instances are the links of Lundin, Elie, and above all St Andrews. In the north-eastern angle of the county is the expanse of Tentsmuir, the largest tract of sand-dunes in Scotland.

Much has been done by man to alter the aspect of Fifeshire. Bogs have been drained and converted into good farming land ; the aboriginal woods have long since disappeared, but hillsides once bare or grassy have been covered with plantations ; and heath-clad moors that afforded shelter to game and made this in Stewart times a royal hunting-ground have been ploughed and cultivated. The name "muir" indeed occurs often in Fife ; but it is the name rather than the reality that survives. Muir of Dysart, Cupar Muir, Magus Muir, Roscobie Muir, are a few chosen at random, two of them celebrated in history.

The Links, St Andrews

And such place-names as Muirfield, Muirton, Muirhead, etc., bear testimony to an earlier aspect of the country.

There is thus nothing strikingly picturesque in Fife scenery : its hills cannot compare in height or grandeur with those of Highland counties, nor do its sea-cliffs, as do those of our wild northern and western shores, strike the beholder with awe. The undulating surface is pleasingly diversified with stream and field and woodland. It is a region where the busy centres of enterprise and activity do not seem greatly to mar the rural amenities of the quiet country-side into which the traveller soon passes. Even its coalfields can scarcely be said to form a "black country." The shore-line though girt in some parts with cliffs is usually low. Here and there it is strewn with huge boulders that have travelled with the ice from the distant hills of the interior in an age long gone by ; and in some places dangerous rocks shelve seawards and skerries protrude from the swelling tide.

5. Rivers and Lakes.

The two rivers whose entire courses lie in Fifeshire are the Eden and the Leven, but there are many smaller streams that merit some notice in a description of the county. Its general contour as sketched in the previous chapter determines the flow of surface water, and with the exception of small, independent rivulets in the East Neuk that find their way directly to the sea, most of the running water flows either into the Tay or the Forth, the Eden or the Leven.

The Eden follows a winding course of 29½ miles through the Howe of Fife and discharges its waters by means of a small, muddy estuary into St Andrews Bay. It has numerous tributaries on both banks. On the left the Motray Water and Moonzie Burn conjoin and enter the estuary together at Guard Bridge about a mile below

The Eden, Cupar

the point where the current of the Eden is arrested by the inflowing tide. Both descend through pretty pastoral valleys from sources in the eastern prolongation of the Ochils, respectively 12½ and 7½ miles distant. On the right bank of the Eden the most considerable affluent is the Ceres Burn, which drains an area of about 25 square miles and joins the Eden near Dairsie. It is formed by the union of a number of smaller streams, the rapid descent of which has caused them to scoop out some of the deep ravines referred to in the last chapter. Dura Den is on the Ceres Burn.

To the south of the Lomonds flows the river Leven, rising just beyond the boundary of the county in Loch Leven and entering Largo Bay at the port of Leven after a course of 16¼ miles. Its main tributary on the right is the Ore or Orr. It rises in the Saline Hills 17 miles from its confluence with the larger stream and winds sluggishly through the coal-field of west Fife. The Eden and the Leven were at one time good salmon rivers, but mill-dams and factories have rendered them valueless in this respect.

The minor streams of the county, inconsiderable when taken separately, carry the largest proportion of drainage to the sea. Their action as erosive agents has been already spoken of. The Kinness Burn and Kenly Water flow into St Andrews Bay down the northern slope of the central watershed. On its other side the Dreel enters the sea at Anstruther, and the Kiel and the Scoonie at Largo Bay. The Bluther Burn, which joins the Forth near Torry, is the largest of a few streams that descend through western Fifeshire to that river.

Evidence, both geological and historical, points to the existence of many lakes in Fife that have now disappeared. These were not in rock-basins like the Highland lochs, but in shallow depressions that have either been silted up by the sediment carried into them by streams, or artificially drained. Their sites are now peat-mosses or alluvial tracts under the plough. It is interesting in this connection to compare with a modern map of the county such an early one as that published in Blaeu's Atlas in 1662. The latter shows, for example, the "Swan Loch" a few miles south-west of Tayport, traces of which can now be found only in local place-names.

Lindores Loch, Newburgh

The lakes that remain, though fairly numerous, are of small extent. In the west are Loch Glow, Loch Fitty, Loch Gelly, Camilla Loch, Otterstone Loch, and Kinghorn Loch. Several of these lochs have diminished greatly in extent from such causes as have led to the disappearance of many of their fellows. Lindores Loch near Newburgh has a circumference of about four miles, and Kilconquhar Loch of two ; and these may have been saved because no burn of any size flows into them to silt up their waters. Good trout-fishing is to be had in many of the streams and lakelets of the county.

6. Geology and Soil.

The rocks are the earliest history books that we have. To those who understand them they tell a fascinating story of the climate, the natural surroundings and the life of a time many millions of years before the foot of man ever trod this globe. They tell of a long succession of strange forms of life, appearing, dominating the world, then vanishing for ever. Yet not without result, for each successive race was higher in the scale of life than those that went before, till man appeared and struggled into the mastery of the world.

The most important group of rocks is that known as *sedimentary*, for they were laid down as sediments under water. On the shores of the sea at the present time we find accumulations of gravel, sand and mud. In the course of time, by pressure and other causes, these deposits will be consolidated into hard rocks, known as conglomerates, sandstones and shales. Far out from shore there is going on a continual rain of the tiny calcareous skeletons of minute sea-animals, which accumulate in a thick ooze on the sea-floor. In time this ooze will harden into a limestone. Thus by watching the processes at work in the world to-day we conclude that the hard rocks that now form the solid land were once soft, unconsolidated deposits on the sea-floor. The sedimentary rocks can generally be recognised easily by their bedded appearance. They are arranged in layers or bands, sometimes in their original horizontal position, but more often tilted to a greater or less extent by subsequent movement in the crust of the earth.

We cannot tell definitely how long it is since any special series of rocks was deposited. But we can say with certainty that one series is older or younger than another. If any group of rocks lies on top of one another then it must have been deposited later, that is it is younger. Occasionally indeed the rocks have been tilted on end or bent to such an extent that this test fails, and then we must have recourse to another and even more important way of finding the relative age of a formation. The remains of animals and plants, known as fossils, are found entombed among the rocks, giving us, as it were, samples of the living organisms that flourished when the rocks were being deposited. Now it has been found that throughout the world the succession of life has been roughly the same, and students of fossils (palaeontologists) can tell by the nature of the fossils obtained what is the relative age of the rocks containing them. This is of very great practical importance, for a single fossil in an unknown country may determine, for example, that coal is likely to be found, or perhaps, that it is utterly useless to dig for coal.

There is another important class of rocks known as

igneous rocks. At the present time we hear reports at intervals of volcanoes becoming active and pouring forth floods of lava. When the lava has solidified it becomes an igneous rock, and many of the igneous rocks of this country have undoubtedly been poured out from volcanoes that were active many years ago. In addition there are igneous rocks—like granite—that never flowed over the surface of the earth as molten streams, but solidified deep down in subterranean recesses, and only became visible when in the lapse of time the rocks above them were worn away. Igneous rocks can generally be recognised by the absence of stratification or bedding.

Sometimes the original nature of the rocks may be altered entirely by subsequent forces acting upon them. Great heat may develop new minerals and change the appearance of the rocks, or mud-stones may be compressed into hard slates, or the rocks may be folded and twisted in the most marvellous manner, and thrust sometimes for miles over another series. Rocks that have been profoundly altered in this way are called *metamorphic* rocks, and such rocks bulk largely in the Scottish Highlands.

The whole succession of the sedimentary rocks is divided into various classes and sub-classes. Resting on the very oldest rocks there is a great group called Primary or Palaeozoic. Next comes the group called Secondary or Mesozoic, then the Tertiary or Cainozoic, and finally a comparatively insignificant group of recent or Post-Tertiary deposits. The Palaeozoic rocks are divided again into systems, and since the rocks of Fifeshire fall entirely under this head, we give below the names of the different systems, the youngest on top.

Palaeozoic Rocks.

Permian System.
Carboniferous System.
Old Red Sandstone System.
Silurian System.
Ordovician System.
Cambrian System.

Few areas of a similar size contain more geological interest than does the county of Fife; within its limits the science may be profitably studied and taught by constant appeals to evidence in the field. Its long coast-line affords almost unrivalled examples of shore-sections; and the interior, which from the point of view of this science is of a most varied character, abundantly exemplifies the characteristic structures of the terrestrial crust.

To begin with the oldest rocks of Fife—those of the Lower Old Red Sandstone—we must visit in imagination the north of the county. Throughout the length and breadth of the Ochil chain and its picturesque eastern continuations that bound the Firth of Tay on the south, the underlying stone belongs to the volcanic series of this deposit. Andesites constitute the great mass of the Ochils, and all may have come from centres of eruption in the south-west where the volcanic series are thickest. Traverses of these hills between the Earn and the Eden, especially along the line of such a great natural opening as Glenfarg, give abundant examples of volcanic agglomerates, tuffs, and intrusive bosses. The imposing rock that rises in perpendicular cliffs above the railway at Newburgh is a huge mass of intrusive andesite, perhaps the grandest in the Ochils. Remarkable shore exposures of andesitic conglomerates occur between Peashill Point

Intrusive Andesite Cliff, Newburgh

and Wormit. These volcanic rocks must have been formed in the great inland sea ("Lake Caledonia") where they became associated with beds of sandstone and even fine argillaceous shale enclosing remains of plants and fishes. This sedimentary zone is overlaid by a series of lavas which constitute the whole of the Upper Old Red

Sandstone from the Tay to the Eden, and which may be studied advantageously on Lucklaw Hill.

To the south of this and roughly commensurate with the valley of the Eden lie the rocks which must be taken next in chronological order. Far down below the surface drifts and deposits of the Howe of Fife lie the red and yellow sandstones and conglomerates of the Upper Red Sandstone. They appear partially to overlap the volcanic series on the north and are abruptly bounded on the south by the Carboniferous rocks and their intrusive masses. In the west they are brought to light in the course of the Glen Burn, a tributary of the Eden that rises between the West Lomond and Bishop's Hill. But by far the most interesting place in which to examine them is Dura Den, which bulks so largely in geological literature for its palaeontological treasures.

The fish-fauna of the Upper Old Red Sandstone have been nowhere found in such perfection as here. In good specimens the dark impressions on the pale yellow stone are very striking. Whole shoals of fish seem to have been killed suddenly; and the marvel is that they have been fossilized before they decayed or fell to pieces, or had their scales and bones scattered. A fine collection of these fossils may be seen in the Museum of St Andrews University.

The Carboniferous system, whether regarded from a geological or an industrial standpoint, is by far the most interesting and important of the rock formations of Fifeshire. The existence and the use of coal were known in this county in very early times—long before the age of scientific geology had dawned. There would at first be no mines, but coal appears at the surface in certain inland water-courses and in many places along the coast of the Firth of Forth ; and it was in such places that this fuel was first found. An early visitor from the Continent noted that the natives burned black stones.

The lowest carboniferous rocks are those of the calcareous sandstone series and in Fife these are closely associated with volcanic material. This has been either contemporaneously erupted and so occurs in alternate successive layers of lava, volcanic ashes, and sedimentary deposits of sea and estuary; or it has been intruded into the strata after their deposition. The former conditions are best illustrated in the west of the county, the latter in the east. The oldest carboniferous rocks yet detected in this district are exposed in the ravine of the Devon and in the banks of neighbouring streams ; and these are supposed to form the eastward extension of the volcanic area of the Campsie Fells ; while the rocks of the Burntisland district, next in point of antiquity, connect with those of Midlothian. The coast section between Kirkcaldy and Burntisland is one of the finest examples in Britain of this alternation of volcanic and sedimentary material. Volcanic intrusion, on the other hand, is best illustrated in the East Neuk. The most complete section of the lower carboniferous rocks is met with on the coast between Pittenweem and Anstruther, where many thin seams of coal varying from a few inches to a couple of feet in thickness are laid bare. These rest on fireclays. Along this coast there has been much faulting and the region is one of intense volcanic disturbance, the igneous rocks having in many cases modified and even destroyed the coal-bearing strata. Though occasionally destroying coal seams, the intrusion of igneous material here and elsewhere in Fife in some cases has had a beneficial influence on them by converting the coal into a better steam coal. The district, too, contains oil-shales which under favourable economical conditions may yet be worked. These shales contain numerous fossil fishes. The calcareous sandstones occupy the largest part of eastern Fife, but are bounded on the north by the line of the Eden valley and terminate suddenly to the west of St Andrews, where we find a stretch of sand flanked by ranges of sand-dunes that prevail northwards to the Firth of Tay.

The upper, and geologically younger, series of the Carboniferous system are the limestones. They occur in the Burntisland district where their strata intermingle with sills of volcanic material in a remarkable manner. Moreover they extend westwards to the county march, eastwards round Largo Bay, and northwards, in an ever narrowing tongue-like area, to within a few miles of St Andrews. A zig-zag line running almost due north from Elie bounds them from the lower carboniferous rocks of the East Neuk. The limestones comprise three series, an upper, a middle, and a lower ; and the middle series, where we find a succession of sandstones, shales, coals, and ironstones, are the seat of the productive coalfields of west Fife, including those of Torryburn, Oakley, Saline, Dunfermline, Kirkcaldy, and Markinch.

The millstone grit, which is superimposed on the carboniferous limestones, consists of coarse sandstone and conglomerate. It is exposed in only one small section of the county between Dysart and Kinglassie, and is especially observable on the shore to the east of Pathhead.

Above this lie the true Coal Measures, which are worked in the Dysart, Wemyss, and Leven coalfields, where there are no fewer than fourteen seams of coal, and to a smaller extent at Kinglassie. The Coal Measures are believed to underlie Largo Bay but to have been cut off by a great fault which lowers them to the carboniferous limestones so that they do not appear any further east.

Being exposed for some seven miles along a continuous natural coast section, the Coal Measures can be better studied in this than in any other district in Scotland. The Dysart Main Coal is the thickest seam in Fifeshire. Some of the coals are mined under the waters of the Firth, and at one point, West Wemyss, workings extend half a mile from the shore. The measures reappear to the south of the Firth in the Midlothian coalfield. It will thus be seen that in Fife coal is derived mainly from two series of the carboniferous rocks: in the west from the middle series of limestones and in the Leven district from the true Coal Measures. A coalfield probably larger than those of Midlothian Kirkcaldy and Dysart put together, lies under the waters of the Forth.

Fifeshire is one of the best districts in Britain for the study of volcanic materials as affecting the various systems of rocks that have been dealt with. Apart from the igneous rocks of the Old Red Sandstone to which reference has already been made, volcanic phenomena may be considered as displayed in sills, bosses, dykes, veins, and necks. Sills, which vary in thickness from an inch or two to several hundreds of feet, as in the Lomond Hills, are igneous material which as molten lava found its way between the planes of stratified rocks and on cooling remained there as a hardened sheet. Though undoubtedly intrusive, this might appear to the uninitiated as one of the strata. Bosses are the outcrop of a pillar or column of lava which has forced its way upwards through the stratifications of earlier rocks. When these are dome-shaped, another name for them is laccolites. Dykes are formed by molten material which has protruded through huge horizontal cracks or fissures, and which being harder than the strata it has penetrated has remained a solid wall whilst the softer materials on either side have suffered more from denudation. Necks may be regarded as the chimneys of volcanoes that have been filled with igneous matter, which has been shot up from a depth of thousands of feet. Smaller intrusions of a character similar to but less regular than the foregoing are known as veins. Injected materials of igneous origin have of course a marked effect upon adjacent layers; they harden shales and sandstones, and in the case of coal-seams by virtue of heat and pressure produce either anthracite or turn them into cinders and even soot, thereby entirely destroying their value.

It is in the west of Fife, and particularly in the Saline

district, that the latest volcanic activity amongst the carboniferous rocks has left the most abundant signs; but the famous Binn of Burntisland, perhaps the site of the actual orifice from which the volcanic material of a wide surrounding neighbourhood was discharged, is undoubtedly one of the most instructive examples in our country of this geologic agency.

The numerous red patches that occur so frequently in the geological map of Fife and particularly in the eastern portion of the county are for the most part *dolerite* sills, phenomena that may be studied to better advantage here than in any part of southern Scotland. The backbone of Fife, formed by the Cleish and Lomond Hills, and trending undeviatingly eastwards to St Andrews Bay, is but a continuation of an intrusive volcanic mass that traverses Scotland from Renfrew to Fife. To the east of the depression crossed by the main railway between Markinch and Kettle, the volcanic range widens out to a breadth of seven miles between Colinsburgh and Pitscottie and then runs eastwards in two ever-diminishing lines which terminate respectively at Kinaldy and Lumbo. Among the sills of Fife must be reckoned the Isle of May, which, though it lies at a distance of five miles from the coast, is both geologically and geographically a part of the county. Many of these sills are probably later in origin than those of the carboniferous rocks. One of the most remarkable dykes in the county has been traced with some slight breaks from Buddon Ness past Auchtermuchty and thence through the Ochils, a distance of 30 miles. Many dykes of less extent but of even more striking appearance occur on the shore and elsewhere in the county.

Volcanic necks, frequent in many other parts of the county, abound around the eastern shores of Fife and in the peninsula between Largo Bay and St Andrews Bay. In this unique district, Sir Archibald Geikie has identified

Rock and Spindle Rock, St Andrews

no fewer than 80 separate vents. Perhaps nowhere in the world can volcanic vents be so advantageously studied. Exposed by denudation, they are found to vary in diameter from a few yards to half a mile. The vent of the Rock and Spindle with its picturesque shore-stack is one of the curiosities of the district. "Rock," as Sir A. Geikie explains, "is here the Scots word for a distaff; and

Spindle has reference to the stellate mass of basalt resembling a spinning-wheel." The volcano-like cones of Carnbee Law and Largo Law, the latter of which, a miniature Vesuvius, is the most conspicuous hill (949 feet high) in east Fife, owe their present shape not to the recent extrusion of igneous material but to the wearing down of that material by ages of denudation. Their original height must have been far greater.

Largo Law

There is no doubt that the whole of Fife was at one time entirely buried under an ice-sheet which, moving first in a south-easterly and then in a slightly north-easterly curve, planed and ground the underlying rocks of the district and brought down from the Central Highlands vast accumulations of drift consisting of boulder-clay which now constitutes in most places the soil of the county. This boulder-clay has penetrated under the waters of the Forth, forming an impervious flooring that may yet render feasible extensive coal-mining operations under the firth. The striae on the ice-worn rock-surface mark unmistakably the trend of the moving ice-sheet; and erratic boulders of rock, familiar in the Ochils and in the Highlands but foreign to their present surroundings, have been left scattered in thousands all over the surface of the district and on the adjacent coasts. The most impressive display of these is to be found in front of Lundin Links. The hollows which occur every here and there throughout the region are due either to scoopings-out in the solid rock or to irregularity of surface left in the drift deposits when the ice melted. Most were at one time filled with water and formed lakes that have either been drained by human skill, or gradually filled up as peat bogs by the growth of vegetation. Thus Loch Glow is a rock-basin, but Loch Leven a sheet of water due to an irregularity of surface. Both kinds date back to the Glacial Period.

To the same agency, too, are due those peculiar mounds of sands and gravels known as kames, remarkable examples of which abound near St Fort in the lower valley of the Motray Water and to the immediate south of Wormit Bay. There is evidence, moreover, of the undoubted fact that the land was at one time at least 100 feet below its present level. This is presented by the so-called Raised Beaches which in three successive terraces of 25, 50, and 100 feet are at certain places found to run parallel with the shore. The terracing of these beaches is best seen near Kincraig Hill. At one part the 100-feet beach penetrates far inland and the skeleton of a seal has even been found within a short distance of Cupar. The 25-feet beach can be traced intermittently all round Fife, but the tract between St Andrews and Tayport is particularly well-marked. The lower parts of many of the sea-coast towns are built on this terrace. Many of the rock-stacks on the shore are relics of the old coast-line. The so-called submerged forests, evidences of which have been found on both the Tay near Flisk and on the Forth near Largo, are somewhat of a geological puzzle. They consist of sheets of peat made up of fresh-water plants, interspersed with remains of trees and are found near low-water mark. These "forests" are not peculiar to Fife nor to Britain, but are found on the continental shores of the North Sea and the Baltic. They are supposed to have grown in littoral marshes or lagoons, which were subsequently submerged.

The soil of any district is of course largely dependent on the character of the underlying rock formations, and hence in Fife it varies considerably in different parts. In the section to the north of the Eden it is quick and fertile, nowhere of great depth but well suited to the cultivation of grass. That which overlies the carboniferous rocks consists for the most part of cold retentive clays, seldom fertile or easily wrought but amenable under modern methods of draining and manuring to such improvement as to secure good crops. The Howe of Fife, or Stratheden, as far inland as Cupar, is very fertile and in parts exceedingly productive. In the parish of Cameron, where the land reaches an elevation of 600 feet, the soil is cold and stiff, clayey and mixed with lime. Around Ladybank it is shingly and has the appearance of having had its richest earthy coating removed by a current of water. Farther south and west it becomes heavier and more valuable; while in the high ground adjoining the Lomonds it is best suited for grass. The valley of the Ore is largely composed of clay, or of thin loam with a strong clayey subsoil, a condition that prevails in the district to the north of Dunfermline; and in the parishes of the extreme south-western reaches of the county we find a mixture of clay and loam that is very fertile. On the coast section from Inverkeithing to Leven it varies from light dry to strong clayey loam capable of superior cultivation. Largo and its environs have a deep rich loam which produces highly satisfactory results in crops. In the neighbourhood of Elie it is light but very fertile. All along the east coast there is clay and rich loam. While rolling stones and outcrops of rock present in this part of the county but little obstacle to ploughing, here and there boulders are met with that have been carried down during the ice-period and deposited on the melting of the ice that must at one time have covered the surface of the land. Near St Andrews the soil is light, and at Leuchars it tends to become sandy, a characteristic that so prevails on Tentsmuir as to render that tract of land almost useless for agriculture. Along the southern slopes of Forgan and Ferry-Port-on-Craig the soil though light and variable is very suitable for farming. On the whole modern methods in farming, fencing, manuring, and especially draining, have within the last half century gone very far to render a district, naturally wet and swampy, one of the most fertile and productive in Scotland.

7. Natural History.

The flora of Britain has been divided into four classes of plants, each adapted to special climatic conditions and named after the region in which these conditions prevail. These regions are: (1) the Alpine, (2) the Sub-Alpine, (3) the Lowland, and (4) the Maritime. It is interesting to trace the steps by which this arrangement of the flora has been established.

Geological evidence and the similarity of the floras of Britain and Western Europe make it practically certain that it was from the latter region that our country was restocked after the desolation of the Ice Age. As the sheet of ice was disappearing, there seem to have occurred two periods of upheaval, during which Britain and the Continent were connected by dry land. Across this land-bridge the vegetation of Europe followed the retreating ice, and reaching Britain gradually filled up the void it had left behind it.

In the van of this advancing army of species, nearest to the ice, came Arctic alpine forms; next came the sub-alpine species; and these were followed successively by the lowland and the littoral kinds of plants. The first two classes readily obtained a footing in the vacant soil of Britain; but, while the two last were still crossing, the land-bridge became submerged. At the next upheaval these forms effected a crossing, and a severe struggle for existence took place between them and their predecessors. The result was that the alpine and sub-alpine species were forced into higher altitudes. Before every species had crossed, the land-bridge was again submerged. The general trend of the advancing species was towards the north-west. Consequently we find that the southern and eastern portions of Britain have a greater variety of species than the north and west; and for the same reason Ireland is poorer in species than Britain.

For variety of species Fife stands second on the roll of Scottish counties. Its hills are not high enough to provide the natural conditions of growth for Arctic alpine plants, but the other three classes are well represented. Along the rivers are moisture-loving plants, on the hill tops grasses and heaths, and near the coast there is a margin occupied by plants adapted to growth by the sea.

The littoral species are sufficiently varied to be sub-divided into five groups: (*a*) Those which grow on the seaward side of the sand-dunes. In this area the characteristic plants are the various seaweeds, and the most abundant of these are the brown laminaria and fucus. (*b*) Plants that grow on the sand-dunes. A typical formation of this sort will be found on Tentsmuir. The first plant to inhabit the loose sand which originally constituted the soil of this region is sea couch grass (*Agropyron junceum*) and sand-sedge (*Carex arenaria*). These plants serve to bind the loose sand together by virtue of their long underground stems, and also by their tenacity of life and the power of their young shoots to push themselves up from below a considerable depth of sand. They are soon followed by lyme grass (*Elymus arenarius*), and the solid ground thus formed gradually becomes more fertile and is colonised by such plants as thistles, docks, and grasses. Higher up on Tentsmuir the dominant species is heather, which is shorter and stumpier the nearer it is to the sea. (*c*) The next class of plants is that growing on the links behind the sand-dunes. (*d*) The fourth, that of plants which grow on muddy ground such as prevails at the estuaries of the Eden and the Tay. *Plantago maritima*, *Armeria maritima*, and *Glaux maritima* are examples. Higher up the Tay at Balmerino may be traced the transition from salt mud plants to fresh-water mud plants. (*e*) Plants which grow on the cliffs. This group includes such forms as the common rock rose (*Helianthemum vulgare*), sea campion (*Silene maritima*), bladder campion (*Silene inflata*), and meadow cranesbill (*Geranium pratense*).

The natural vegetation of the lowland region in Fife has been largely defaced by cultivation, but can still be studied in the dens, glens, and woods. The trees of the county are the oak, found where the soil is rich and deep; the Scots pine, which grows in poorer soil and moderately drained peat—for example on Tentsmuir; the birch, which thrives in opener and more exposed parts; and alders and willows, which are found along the margins of rivers. There are many noble representatives of the beech in Fife, but it is not an indigenous type. It was introduced from south Britain and the Continent, and has proved a serious competitor to the oak in lowland woods. Its power of adaptation to many soils and its shade-producing character has in many cases transformed the oakwoods, especially with regard to their undergrowth. The beauty of the opening woods is enhanced by such flowers as the violet, primrose, and hyacinth, and in beechwoods the wood sorrel (*Oxalis acetosella*) frequently forms a green carpet. There are some celebrated specimens of trees in Fife. At Otterstone and Donibristle are two ash trees; there are oaks, Spanish chestnuts and sycamores at Aberdour, Donibristle, and Balmerino; and yews at Forgan. Some of these attain a height of 80 feet and a girth of 20 feet, while the diameter of their branches is in one case 72 feet. The lowland region also includes many lakes and ponds in which a great variety of water plants flourishes.

The sub-alpine region is best represented on the Lomonds and the Ochils. The Lomonds are characterised by large tracts of moorland; but on the Ochils this is not the case: they are noteworthy for their greenness. On the moorland areas of the Lomonds flourish in abundance heather and blaeberry. The bogs here also are typical and their vegetation consists of peat-forming and peat-frequenting plants, such as bog moss (*Sphagnum*), polytrichum, crowberry, butterwort, sundews, and bog-myrtle. Some rare forms such as bird's-nest orchis (*Neottia nidus avis*), the filmy fern (*Hymenophyllum*), and the parsley fern (*Cryptogramme crispa*) are found on the Lomonds and Knockhill. The coral root orchid is found on Tentsmuir.

Yew Trees, at Forgan

What has been said as to the return of plant life to Britain after the Ice Age applies equally to animal life: whilst the vegetation of the Continent was crossing it was accompanied by the fauna of the districts from which it came, and most of the species we possess are also found in continental Europe. The variety of our fauna, moreover, was restricted in the same way as that of our flora; with this important difference that even after the submergence of the land-bridge, forms that could take an aerial or an aquatic course are found in this country in as great variety as in their original home; and, indeed, it is amongst the birds, insects, and fishes that species peculiar to Britain occur. In the same way we find that Ireland has fewer species of mammals, reptiles, etc. than Britain, while its bird and insect life is as rich as our own. Thus Germany has 90 species of mammals and Scandinavia 60, but Britain has only 40; and the nearest continental country has 22 species of reptiles and amphibia as compared with Britain's 13 species. We have, however, the testimony of the geologist to the fact that such creatures as the bear, the rhinoceros, the tiger, and the hyena haunted our primeval forests; and after the advent of man, though in prehistoric times, there is evidence of the presence in this country of gigantic animals like the *Bos primigenius* or *Urus*, large as elephants, but swift and fierce.

Foxes, hares, rabbits, squirrels, weasels are commonly spoken of as wild, but these are animals of cultivation. Twenty-five species of mammalia, including the above, are found in Fife. The otter is occasionally met with along the banks of the Eden, but there is a tendency for this animal to become extinct. Badgers used to be found on the Kinkell Braes at St Andrews, but have been exterminated. The county is an excellent one for fox-hunting, and the Fife Foxhounds have frequent "meets," particularly in the hilly parts where excellent cover is afforded in the woodlands and fields. The marten was taken a number of years ago near Falkland; and there used to be pole cats. Hedgehogs and voles are common, and so too are the rabbit and common hare, though the mountain hare does not appear to come so far east. The common rat seems to have exterminated the black rat. The common seal and the grey seal are plentiful in the estuaries of the Tay and Forth; and porpoises and even whales are not unknown. The first recorded Scottish example of a hump-backed whale (*Megaptera longimana*) was taken off the mouth of the Tay in 1883, and a specimen of Sowerby's whale was cast up at St Andrews. Reptiles and amphibians of the Tay basin include the adder, slow-worm, viviparous lizard, newt, frog, and toad.

The marine fauna of Fife do not differ materially from those of the east coast in general, and although there are fewer species than on the west coast, where the warmer waters of the Atlantic foster a rich marine life, several extremely interesting and uncommon forms have been secured in St Andrews Bay and the Firth of Forth.

Among the mollusca mention must be made of the fresh-water bivalves of the Tay—the prey of the pearl fishers. The group Coelenterata is also exceedingly well represented in the Firth of Forth, nearly all the species known to occur in Scottish waters being found here.

Two very interesting species of the sea anemone have been taken from the Firth of Forth. One of these, commonly known to naturalists as "granny," was taken by Sir John Dallyell from the Firth in 1828, and lived until 1887. Another species of interest is the plumose anemone—the finest British species of sea anemone. No fewer than 12 different species of sponges are also known to exist in this firth.

There are at least 156 species of birds found in the ornithological district to which Fife belongs. These have been classified as (1) resident, (2) summer migrants, (3) winter migrants, (4) those that occur only during the spring or autumn migration or both, but that do not stay in the locality, (5) those of occasional occurrence, and (6) abnormal. The lark nests on field and hill and adds to the pleasure of the golfer on the breezy links. The missel thrush pipes out his tuneful lay in the intervals of the March gales; later the mavis gladdens the spring morning by his cheerful matin; and the blackbird's mellow vesper enhances the calm beauty of evening. The most prominent singer in the early summer months is the chaffinch. In late spring the summer migrants assemble once more in our land. These are the warblers, the swallows, the cuckoo, the corncrake. The wood pigeon arrives every autumn in flocks from Scandinavia. The eider duck breeds on Tentsmuir, and here too the red grouse and Pallas' sand grouse are found. The king duck is a regular winter visitant to the Tay and St Andrews Bay. In the same region the turnstone breeds. The stone chat and the Sandwich tern appear in summer. Amongst winter migrants are the golden eye, the velvet scoter, and the mountain linnet. The storm petrel is sometimes blown inwards in violent weather. Amongst occasional visitants have been noted the great grey shrike, the hawfinch, the little auk, various kinds of skua and the osprey; while birds abnormal but reported as seen in Fife include the little bustard, the white and the yellow wagtail, the shore lark, the hoopoe, the bittern, the stone curlew, and, rarest of all in eastern Scotland, the black redstart and the white spotted blue-throat.

Amongst special haunts of birds in Fife may be noted the two estuaries, the lakes, the Isle of May, and Tentsmuir.

A county with so rich and varied a flora as Fife cannot fail to have a large and representative host of insects. A species of louse (*Platyarthus Hoffmanseggii*), which inhabits the nest of ants, is found in Scotland only in Banff and Fife.

8. Round the Coast — Newburgh to Fife Ness.

The part of the coast-line to be traced in this chapter may be divided into two parts, the river frontage from Newburgh to Tayport, 20¾ miles long, and the section exposed to the sea between Tentsmuir Point and Fife Ness, which measures 14½ miles in a straight line but is 24 miles along its curvature. The estuary consists of intricate channels between long banks of sand and alluvial mud brought down by the river, which however deposits most of the sediment on its left bank. Hence the navigable channel is on the Fifeshire side.

Opposite Newburgh is Mugdrum, a narrow flat islet nearly a mile long. Above this the river rapidly narrows, though the tide flows up to a point two miles beyond Perth. The stretch between Perth and Newburgh is famous for its salmon fisheries, the most productive in Scotland.

Newburgh, a prettily situated township dominated by a striking and precipitous crag of volcanic origin, is not so new as its name would indicate. It was founded as a burgh of barony by Alexander III in 1266, a "novus burgus juxta monasterium de Lindores."

Farther down the river on either side of Flisk, geologists have discovered an interesting example of the so-called "submarine forest" already referred to.

Below Flisk is the fishing village of Balmerino, with a ruined abbey built by the queen of William the Lion. Between Balmerino and Newport cliffs of andesitic conglomerates face the river except on the small bay of Wormit. Peashill Point at the western end of its curve is of special interest geologically, and at the east end the Tay Bridge has been constructed across the firth to Dundee. The present structure, one of the longest in the world, superseded the first Tay Bridge which was destroyed during a terrific gale in December, 1879. The new bridge, which carries a double line of rails, forms one of the main links in the east coast route between London and the north.

New Tay Viaduct, from the North

Wormit and the older villages of West and East Newport now form the burgh of Newport. Its elegant villas are arranged for the most part in terraces parallel to the river, which afford exquisite views of the broad expanse of the estuary, the Carse of Gowrie, Dundee, and the Sidlaws. The wooded and grassy heights behind Newport are very picturesque.

The Tay is here nearly two miles broad, but it deepens and narrows to less than a mile between Broughty Ferry and Tayport, about three miles farther east. A difficult passage flanked by Barry Sands on the north and the Abertay Sands on the south must be followed for nearly 10 miles before the open sea is reached. The channel from Dundee outwards is regularly dredged and carefully buoyed and charted, so that the port is accessible to the largest steamers. A lightship marks the entrance of the channel opposite the end of the long spit of the Abertay Sands. Tayport, formerly known as Ferry-Port-on-Craig, is the seat of an ancient ferry dominated by a rock or craig, whence the name. From Tentsmuir Point southwards to the mouth of the Eden, a distance of about seven miles, the coast is low and sandy, nowhere attaining a greater elevation than 16 feet, and entirely destitute of even the smallest port or village. Near the mouth of the Eden is Guard Bridge, with an important paper-mill. Proceeding farther along the shore we reach Pilmour or St Andrews links, so famous for golf.

St Andrews stands on one of the raised beaches of the county, about 50 feet above sea-level. Long, shelving rocks strike out seawards, the scene of many a shipwreck during winter storms. Its small harbour is difficult and often dangerous to enter, and apart from the local export of grain and potatoes and the import of coal, timber, guano, salt, and slates, its industry so far as the port is concerned is confined to fishing. There is, however, a considerable herring-fleet, and in the quarters that adjoin the harbour a fairly large population of fisher folk.

The Caves, Crail

The rock-bound coast from St Andrews to Fife Ness, a distance of some 12 miles, is replete with geological interest. Its caves, like that of Kinkell, and its shore stacks, such as the Maiden Rock and more particularly the famous Rock and Spindle, are often visited. It is skirted by cliffs from 30 to 40 feet in height. The Ness, which forms the most easterly point in Fife, is a low rocky headland. The Danes' Dyke, said to have been built in 874, runs across it. A mile to the north-east are the Carr Rocks, a dangerous reef now marked by a lighthouse. It has been the scene of many a wreck. In 1844 the *Windsor Castle*, a passenger steamship, was dashed to pieces on it; and during the decade that ended in 1881 no fewer than seven vessels came to grief near the same spot.

9. Round the Coast — Fife Ness to Kincardine-on-Forth.

Commercially and geologically the coast of the Forth is far more important than that described in the preceding chapter. In general character it is rocky with occasional commanding cliffs. Grassy banks from 10 to 60 feet high prevail from Fife Ness to Pittenweem, where the carboniferous strata form a cliff. Shore stacks and gullies abound. At Kincraig, a mile west of Elie, there is the most conspicuous cliff on this part of the coast. It is a precipice 200 feet high and two-thirds of a mile long, and with the adjacent caves and chasms constitutes a striking piece of coast scenery. To the west of this cliff occur conspicuous relics of the raised beaches, where three terraces are distinctly recognisable. The shore of Largo Bay ($6\frac{5}{8}$ miles in breadth and $2\frac{1}{4}$ in depth) is for the most part low and sandy, though at Lundin and elsewhere great boulders have been deposited at the close of the Ice Period. Striking shore sections referred to in the chapter on geology occur all along the northern coast of the

Firth. At Earlsferry and Lundin there are excellent golf-links.

This open, seaward stretch of coast, which extends 20 miles from Fife Ness to Leven, has numerous old-world burghs every here and there along its entire length —quaint, red-tiled fishing villages or little sea-ports, whose houses cluster round a grey church spire. Most of these have at one time or another done considerable trade with Holland, Norway, and other continental countries, and at several are ruined castles that bespeak an importance to which they can no longer lay claim. In our own day the main industry of these towns is fishing and the various handicrafts connected with it. The centre of this industry is Anstruther, which is the head of a fishery district that is neighboured by those of Leith and Montrose. Between it and Fife Ness lies Crail, and to the west Pittenweem ("the town of the cave"), St Monans, Elie, Earlsferry, and Largo.

From Leven to North Queensferry the coast industries are not limited to fishing: coal is the chief export from such ports as Leven, Methil, and Burntisland. The names of East and West Wemyss, also small coal-ports, recall the farther east Pittenweem, in all of which the word that means "cave" seems to be enshrined. All along this coast-line caves abound, and one near Dysart has the repute of being the original dwelling of St Serf, who first brought Christianity into Fifeshire. By far the most important coast town and port in the county is "the lang toon o' Kirkcaldy," with which Dysart, though a separate burgh with a harbour of its own, is now virtually linked by street and tramway. Three miles south of Kirkcaldy is Kinghorn, with a small harbour. It was at Kinghorn Ness, near Pettycur, the highest part of the coast in this neighbourhood, that King Alexander III lost his life while riding one gloomy night along the coast (1286). The precipice over which he was pitched by his horse, is known as King Alexander's Crag, and the spot at which the accident happened is marked by a monument.

Many a winding bay diversifies the coast-line between Pettycur and North Queensferry. The importance of Burntisland as a port has been already mentioned. In the days before the building of the Forth Bridge there was a steam-ferry between that town and Granton, near Edinburgh.

Aberdour, three miles west, with its wooded heights and pleasant shore, attracts many visitors. Its parish includes the island of Inchcolm, as does Kinghorn that of Inchkeith. Aberdour is indissolubly associated with the grand old ballad of "Sir Patrick Spens." Inverkeithing, an old-world royal burgh, where Queen Annabella, mother of "the poet king" lived and died, was anciently of much more importance than it is now. Its small harbour can be used to advantage only at spring-tides.

Between South Queensferry and North Queensferry the great Forth Bridge has been erected, one of the engineering wonders of the world. Its northern end is on Ferryhill Peninsula, whose projection into the Forth, and the fact that the small rocky island of Inchgarvie lies directly in a line with a gap in the southern shore, made its construction feasible. The length of the structure,

Forth Bridge

including approach viaducts, is one mile 972 yards; its greatest height above high water is 361 feet; it affords a clear headway at high water of 150 feet; and its deepest foundation is 80 feet below high water. The total cost was £3,000,000. From 3000 to 5000 men were employed during its construction; and a staff of about 50 men are constantly at work in connection with it. Three years are spent in the painting of it; and the average yearly cost of maintenance is nearly £5000.

Near the north end of the bridge is St Margaret's Hope, a small bay on the shore of which Malcolm Canmore's bride landed on her way to Dunfermline. Hard by stands the ancient castle of Rosyth on a bay selected by the Government in 1903 as a naval port and base. In the adjoining lands nearly 1500 acres have been purchased at a cost of £122,500; and the local authority have been advised that within 20 years there may be a population of 30,000.

A little to the west of Rosyth are Limekilns and Charlestown, adjoining the Earl of Elgin's mansion of Broomhall. They are practically one, and form an outlet for the lime, iron, and coal produced on the estate; but much of their shipping has disappeared with altered modes of transport. At the head of Torry Bay, a bow-shaped bend on the river, the shore of which is here low and flat, are Torryburn and Culross. The former was at one time regarded as the port of Dunfermline, but has now little trade. The latter retains but few signs of its ancient importance. It once did a large trade in salt and coal, and as many as 170 foreign vessels are reported to have

St Margaret's Hope, showing Rosyth Castle and Site of New Naval Base

lain in the offing simultaneously awaiting cargoes. But these days are past, and Culross is now a sequestered riverside village with a story of its own.

Our long journey round the shores of Fife ends at Kincardine-on-Forth (the suffix is necessary to distinguish it from half-a-dozen other places of the same name in Scotland). It has a good quay and roadstead where 100 vessels could ride in safety. It used to have considerable shipbuilding, and in days prior to the introduction of railways its ferry was a place of constant resort for travellers.

The estuary of the Forth differs from that of the Tay in a few important respects. It is deeper. Under the Bridge, for instance, the water in the centre of the channel has a depth of 35 fathoms; whereas between Broughty Ferry and Tayport the trough of the river is only 13 fathoms. Hence in addition to the numerous ports above referred to, we have many others like Grangemouth, Bo'ness, and Granton on the southern shores of the Forth. It is broader. Between Fife Ness and Tantallon Castle the distance is 15 miles, and although it narrows to about one-half of that opposite Elie, it again expands to an equal extent at Leven. Its most important ferry—Burntisland to Granton—is seven miles wide. Indeed, the part of the Forth above the Bridge has dimensions similar to those of the entire estuary of the Tay. Lastly, it is much more exposed to the action of the sea, with the result that there are no such alluvial deposits here as in the smaller estuary. This affects not merely its commerce, but the entire character of its outer shores.

10. Coastal Gains and Losses.

We talk of the everlasting hills and may think, as we gaze at the sea breaking apparently in vain against some rock-bound coast, that its cliffs too are everlasting, but in neither case is the use of the word correct : the hills and the sea-shore are alike vulnerable and are constantly undergoing sure, if slow, change. On the other hand, the raised beaches, of which we have seen remarkable and abundant traces all round the shores of Fife, give evidence that the land has been subject to great successive upheavals in past geological ages, and that in this way it has gained at the expense of the sea.

On the Fifeshire side, at the mouth of the Tay, the stretch of mud and sand that extends for more than six miles seawards from Tayport is a good example of what the combined action of river and sea is doing to build up what may yet become solid land. Between Tentsmuir Point and the mouth of the river Eden, and to the south along the West Sands of St Andrews, we have on a still more extensive scale an example of coastal gain. But these are the chief instances in the county of this phenomenon. Other and smaller accumulations of marine alluvium are met with at intervals on the coast from St Andrews, round Fife Ness, and on the northern shores of the Forth as far as Largo Bay and Leven. At Kincardine-on-Forth, in 1823 and 1829, two embankments were made, at a cost severally of £6104 and £14,000, and by their means 366 acres have been added to the foreshore.

But on the whole it must be said that the story of the Fifeshire coast, as of the British shores in general, is a story of loss. And it is the constant action of the waves and their terrible violence during storms that play the most important part in coast erosion. If the material on which they act is at all soft—boulder clay, for instance—the effect is all the more striking; and often after a storm a cornice of turf is left overhanging the shore, the subsoil having been washed away. The very shingle which has accumulated on the beach becomes a weapon of destruction for the waves to use. This is driven with the force of a battering-ram against the lower part of a cliff, which gradually begins to overhang the shore and in time topples down on it. On the other hand, the erratic blocks, which as we have seen have been left in such abundance on certain parts of the coast at the close of the Ice Age, being of exceedingly hard material, act as a breakwater against the fury of the waves. This may be well seen at Anstruther and Cellardyke.

Let us now seek for some historical evidence of coastal loss in Fife. At St Andrews part of the cliff on which the castle is built gave way (1801), and a section of the wall that enclosed the court-yard fell with it. To prevent as far as possible a recurrence of this, great masses of masonry have been built against the face of the precipice, but even these require constant attention and repair. In Sibbald's History of Fife (1710) we read of a tradition "that the ancient Culdees, Regulus and his companions had a cell dedicated to the Blessed Virgin, about a bow-flight to the east of the shoar of St Andrews a little without the end of the peer upon a rock called at this day Our Lady's Craig: the rock is well known and seen every day at low water. The Culdees thereafter, upon the sea's encroaching, built another house where the house of the Kirkheugh now stands, called Sancta Maria de Rupe, with St Rule's chapel." The same authority relates that in his day there lived people in St Andrews who remembered to have seen

St Andrews Castle

men play at bowls upon the east and north sides of the Castle. These places are now covered by the tide.

The farm of St Nicholas adjoining the East Sands has suffered from the inroads of the sea, and the foot-path once ran much nearer the water than it now does.

"The Prior's Croft" is the name given to a part of the shore at Crail from the fact that it at one time formed the gardens of a small religious house, the remains of which now adjoin the very beach.

A century ago at St Monans, grassy knolls are known to have covered ground that is now well within high water mark. Near by at Newark Castle the foundation of an arch remains on the cliff, but the arch itself has disappeared. Before the seventeenth century the links of Burntisland, now limited to a narrow strip along the shore, are said to have extended to the Black Rocks, half a mile distant in the waters of the Forth.

11. The Coast—Sandbanks and Lighthouses.

The navigation of the Forth, and still more that of the Tay, is difficult, and pilots must be employed to guide ships up the tortuous passages that lead to the ports on these firths and avoid the numerous sandbanks and rocks that beset their channels. The sea-coast and its environs, too, are very dangerous—how dangerous the following statistics will serve to show. During the decade that closed in 1908, no fewer than 87 vessels of various kinds were stranded on sands and rocks in this neighbourhood; and reference to the special wreck chart published by the Board of Trade reveals the fact that in the year 1907—8, 31 casualties of a more or less serious character occurred in the same locality. These facts eloquently bespeak the need of having points of danger well marked by buoys, beacons, lightships, and lighthouses.

Although towers and beacons have from almost time immemorial been erected here and there on our coasts, it is only within the last century that thorough attention has been paid to the subject. Thus in 1635 Alexander Cunningham obtained from Charles I a charter for the Isle of May with permission to erect a lighthouse on it. He erected a tower 40 feet high on the top of which a fire of coal was constantly kept burning. The Commissioners of Northern Lights purchased the island for £60,000, and in 1816 erected the present lighthouse on it.

The Admiralty List of Lights now contains a register of 1257 lights on the coasts of the British Islands. The central authority is known as the Elder Brethren of Trinity House, London; but the lights of the district we are considering are under the charge of the Northern Lighthouse Board, Edinburgh. An ample revenue for the maintenance of lighthouses is obtained from light-dues levied on shipping. Out of this the authorities erect and maintain lighthouses, lightships, beacons, buoys, etc. It is their duty also to appoint and license pilots, and remove wrecks that may endanger navigation.

There are 45 lights round the Fife coast. The Tay Bridge is lighted on both sides in the middle of the river where the channel lies. Two lighthouses are erected a little to the west of Tayport; and there is a pile lighthouse with a fogbell a little to the south-east. The mouth of the river is lighted on the Forfarshire side at Buddon Ness by two lamps of 7000 candle power each. Off Abertay Spit a lightship is anchored in 5½ fathoms of water. It is coloured red, has two masts, and the word *Abertay* painted on its sides. The Bell Rock (392,000 candle power) and the May Island lights flash across St Andrews Bay to each other, and are visible respectively at distances of 15 and 21 miles. They are two of the most powerful on the British coasts. The May has also a fog siren. Between them near Fife Ness is the North Carr light-vessel, with a red conical cage containing the light, and a siren that in foggy weather gives forth two

The Abertay Lightship

piercing blasts in quick succession every two minutes. Elie Ness flashes a white light of 2000 candle power every six seconds. Inchkeith has a light of 167,000 candle power at a height of 220 feet above high water mark, and a fog siren. Oxcars Rock, Inchgarvie, and Beamer Rock, are each well lighted; and at the end of each cantilever on each side of the Forth Bridge is a fixed white light to define both the north and the south navigable channel.

The Inchkeith lighthouse was erected in 1804, the Bell Rock in 1811, the May in 1816, the North Carr in 1887.

Sailors about to enter the Tay have the Bell Rock and the Carr Rock lights on either side of them, and similarly the Bass Rock, on the Haddingtonshire coast, and the May lights mark the wide entrance of the Forth. These four stretch in an irregular line between North Berwick and Arbroath at fairly equal distances apart.

These are the outstanding lighthouses of the Fife coast, but besides, every port and every pierhead has its distinguishing light, all of which are carefully charted.

12. Climate and Rainfall.

By the climate of a district is meant its average weather. This is largely determined by its temperature and its rainfall, and very much depends on its altitude and its distance from, or proximity to, the sea. No British county is large enough to have a perfectly distinctive climate of its own; each shares in that of the geographical region to which it belongs.

If its position on the globe had alone to be considered, Scotland should have the climate of Labrador, a country which has a similar latitude, but we know that this is not so: the climate of Labrador is extreme, that of Scotland temperate. This is mainly due to the fact that the Atlantic Ocean which washes its western shores stores up the heat imparted to it in summer and gives off this heat during the cold winter months. The prevailing south-westerly winds blowing off the Atlantic convey this warmth to Scotland, so that in winter we find the warmest zones in the west. To a less degree the North Sea is a repository of heat, and winds blowing from it on our east coast impart a warmth that does not penetrate far inland. Hence, if we look at an isothermal map showing zones of equal heat for the month of January, we shall find that along the west there runs the isotherm of 40° Fahr. in a line roughly parallel to a meridian, farther in this sinks to 39°, in the centre it is 38°, but it rises slightly in a strip that flanks the east coast.

On the other hand, the July isotherms decrease from east to west. That along the west is now 57°; farther inland it is 58°; and a line curving inland from Aberdeen and reaching the Solway marks off a district in central and south-eastern Scotland that has an average temperature of 59° instead of the 38° of winter. Now it is found that the healthiest temperatures are those which range from 50° to 60°. Scotland is therefore climatically a healthy country.

It will be seen on the maps referred to that the Fife peninsula barely touches in winter the isotherm within which the average temperature is 38°, and that in summer it lies entirely between the isotherms of 59° and 60°.

Oban, which lies directly west from the Firth of Tay, is situated on the winter isotherm of 39°, and midway between those of 57° and 58° in summer; its average temperature is therefore slightly warmer in winter and slightly cooler in summer than that of St Andrews.

Generally speaking the climate of Fife is more bracing than that of the west coast.

The following particulars, culled from papers published by the late Dr Buchan in the Journal of the Scottish Meteorological Society, will be of interest. During the period of 40 years ending in December, 1895, the highest mean annual temperature recorded in Fifeshire was that of St Andrews, viz. 47·1°, and the lowest that of Feddinch Mains, viz. 46·0°. Taking mean monthly temperatures, the highest records were those of Burntisland and Leven—58·2°; and the lowest that of Feddinch Mains—36·6°, where also the greatest range was observed, viz. from 36·6° in January to 57·7° in July. At St Andrews the difference between January and July was 19·8°, and at Leven 20·9°. Records kept on the Isle of May showed a range from 39·0° in January to 56·7° in July.

The cultivation of a country does a great deal to modify its climate. Much of the surface of Fife was at one time covered with lakes and bogs. But within the two last centuries these have been drained, and the reclaimed land has done much to improve the climate of the county. The numerous plantations that have sprung up in many places have done their own part in sheltering the land from the biting east winds which prevail in the spring, and early summer months. Over these swamps and lakes, too, noisome fogs were wont to hang; and their disappearance after the reclamation of the land has promoted the health of the inhabitants. Fogs and hoar frosts, however, are not infrequent, and the latter do considerable damage to the root crops.

Localities that have a northern exposure have a slightly more rigorous climate than those that face the south. Thus the snow line is lower on the Swiss than on the Italian side of the Alps. Similarly, though of course in a much smaller degree, land in Fife that slopes to the Forth must be sunnier and more genial than that which slopes to the Tay. The character of the soil and that of the vegetation have also an influence: a sandy soil is more liable to extremes than a heavy clay soil; and winds are tempered by belts and clumps of woodland.

Rainfall is affected by temperature. As soon as the temperature of the air falls below the point of saturation the clouds dissolve in rain, or, if the lowering is very great, in sleet and snow. The character of the winds as regards moisture or dryness and the configuration of the land surface are important determinants in the matter of rain. Now, speaking of Scotland generally, the prevailing south-westerly winds from the Atlantic come to our shores laden with moisture, and as they are proceeding from a warmer to a colder latitude the clouds they convey condense in rain and the west is wet. This is increased by the fact that the mountain systems of the country prevail in the west, and their heights condense the moisture that is in the air. Where, however, the country is opener, as it is in the valley of the Clyde and Forth, more clouds are brought eastwards, and the counties which lie in that region are more liable to rain than where the moisture has spent itself farther west. This applies to some extent to the west of Fife. On the other hand, north-easterly winds are very often rainless; and hence the East Neuk

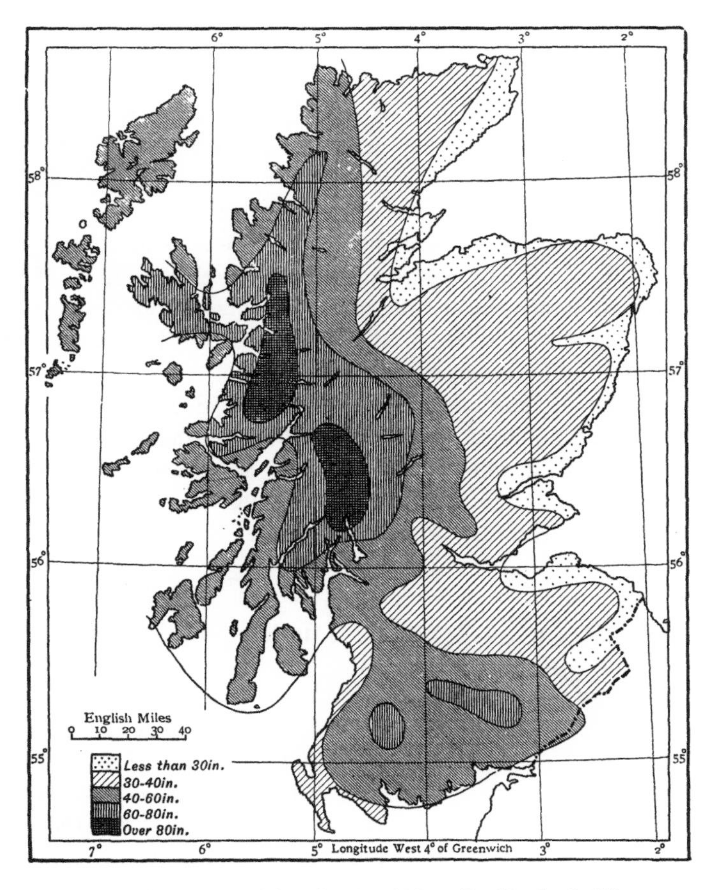

Rainfall Map of Scotland. (After Dr H. R. Mill)

is comparatively dry. A line drawn straight west from Fife Ness to Loch Linnhe would pass through four zones whose rainfall in order from east to west is respectively 25 to 30, 30 to 40, 40 to 60, and 60 to 80 inches annually, while in some spots, as near Tyndrum, more than 80 inches are attained. Fife, divided by a line joining the mouths of the Leven and the Eden, lies within the two first of these zones. It should, however, be noted that the rains attendant on thunderstorms are characteristic of the east rather than of the west, and these heavy summer rains raise the average fall in the east. They also do great damage to crops.

The highest mean annual rainfall in Fife for the period of 29 years ending with 1898 was that gauged at Lothrie Reservoir—39·36 inches; and during the 28 years up to 1898 that of Cupar, 27·30 inches, was the lowest on record.

13. People—Race, Dialect, Population.

At the dawn of history the peninsula seems to have been inhabited by a branch of the Celtic race, the Southern Picts. The Boresti, or Horesti, among whom, as Tacitus tells us, Agricola led his army after the battle of Mons Graupius, have been regarded as dwelling in this region.

Dr W. F. Skene has shown that immigrations of Frisians began about the middle of the fourth century, and that these intruders settled along the firths of Forth and Tay; and it has even been asserted that Frisian peculiarities of idiom are found in these regions, and there alone.

It was amongst the Celts that the early apostles of Christianity in Scotland, St Serf, his coadjutors and successors, commenced their good work. The Saint's name is associated with Dysart, Culross, Loch Leven, and other places in the west of the peninsula. Under his influence the Picts of that region became Christians. St Andrews was another missionary centre; and religion had a large share in the union of the tribes of northern Scotland.

The Danes' Dyke, near Fife Ness, reminds us of the inroads of the Northmen, many of whom, as names of numerous homesteads show, settled in this part of the country.

Soon after, if not even before, the union of Scotland under Malcolm II, the Angles of the Lothians must have had intimate dealings with their Celtic neighbours to the north of the Forth; and southern influence would certainly spread rapidly into the royal county when Malcolm Canmore married the Saxon Margaret. The king himself had spent fifteen years of exile in England, and, Celt though he was, came to like Saxon speech and Saxon ways more than those to which he was native. These predilections were for long the source of sufficient jealousy and trouble amongst his subjects, but they were bound to have a marked and lasting influence on the people of Fife.

Malcolm Canmore (d. 1093) seems to have been the last of the Scottish kings to speak Gaelic; and that language must rapidly have died out in Fife after his day. At least under his immediate successors, and especially in the time of David I, Norman nobles were put in possession of most of the land in the county. There are many Celtic place-names within its borders, but few Celtic or Saxon family names. On the other hand, Norman names are very numerous: Melvilles, Lundins or Lundies, Oliphants, are early instances of Norman families who held lands in Fife.

In the newly established towns trade was fostered and enjoyed privileges which induced foreign settlers to immigrate. Not the least important of these were the Flemings, who brought with them skill in manufactures. Maynard, a Fleming, was the first provost of St Andrews. "As far as I have been able to learn," wrote Dr Harry Spens in 1778, "the original inhabitants of Buckhaven were from the Netherlands about the time of Philip II. Their vessel had been stranded on the shore. They proposed to settle and remain. The family of Wemyss gave them permission. They accordingly settled at Buckhaven. By degrees they acquired our language and adopted our dress, and for these three score years past they have had the character of a sober and sensible and industrious and honest set of people." These sentences indicate a kind of immigration—and the foregoing is probably no singular instance—that has done much for the industry of Fife, and not a little to mould the character and habits of its modern population. Nor, in speaking of foreign influence in Fife, can we overlook that of our ancient allies, the French.

There is a tradition that certain shipwrecked mariners of the Armada settled in Fife, and to this has been traced, somewhat fancifully perhaps, the dark complexion of some of the modern inhabitants of Anstruther, where the Spaniards are said to have settled.

The foregoing review of changes in people and customs that have swept over Fifeshire within historical times must not lead us to forget that after all the county is self-contained beyond most others in Britain. Shut in by the Ochils, the Firths, and the sea, the "Fifers," as by a kindly pleasantry they are often called, occupy a land peculiarly their own; and their characteristics may be attributed largely to that isolation. Along with this must be considered their ancient importance as the inhabitants of a county that was the chosen home of our kings; the large number of royal burghs within their bounds; the somewhat minute sub-divisions of land that multiplied the number of their lairds; and last, but not least, the abundant variety of their callings. Hence their thrift, their independence, their shrewdness, their pawky wit, their love of argument.

A few generations ago the Lowlands of Scotland were largely bi-lingual: English was universally understood and used as the polite tongue and the language of literature; but the vernacular was—and still lingeringly remains—the dialect known as Braid Scots. Under modern conditions this vernacular is rapidly dying out.

The census of 1901 showed the population of Fife to be 218,840. In 1801 there were 93,743 persons within the county, so that within the century the population was more than doubled. Rather more than half of the increase took place within the latter half of the period, and that in spite of the fact that between 1851 and 1861 there was an abnormally small number added. In recent times fewer hands have come to be needed on farms, and hence many country people have either emigrated or gone into the towns. Nearly three-fourths of the people of Fife live in the towns and villages and the remainder in rural districts. Of the 15 towns no fewer than 11 are situated in the southern part of the county, St Andrews (7621) and Cupar (4511) being the only fairly large centres of population to the north of the Lomonds. Kirkcaldy (34,079) and Dunfermline (26,600) are the two largest towns in Fife. The number of females exceeds that of males by some 10,000. The number of people to the square mile in Fife is 434, against 1524 for Lanarkshire and 151 for Scotland.

14. Agriculture. Main Cultivations, Woodlands, Stock.

In few counties is there greater variety of occupation than in Fifeshire: the crafts of the farmer, the miner, the weaver, and the fisherman are generously represented within its area.

From the days of the monks onward Fife has been assiduously cultivated, and some of the most productive soil is found on the borders of the county. Progress in farming, especially during the last century, has been made by leaps and bounds. Thanks chiefly to the farmer, the rental of the county was more than trebled between 1800 and 1900. Landowners and their tenants alike by dint of draining, fencing and manuring, and by the adoption of the most recent agricultural appliances, have made Fifeshire one of the best farming counties in Scotland.

Fife stands sixteenth in size on the list of Scottish counties, its acreage being 322,844 as against 2,695,094 for Inverness, the largest, and 34,927 for Clackmannan, the smallest. Its farmers have for several generations devoted their best skill and attention to the cultivation of cereals. Of this area, 171,075 acres are under cultivation, and the latest returns (1909) of the Board of Agriculture show Fife as the premier Scottish county for the production of wheat—10,686 acres. Oats are however its largest crop: 38,876 acres are devoted to this; while barley accounts for 19,885. Beans, peas and rye are of less account.

Potatoes and turnips are the chief green crops: of the former there are 16,645, and of the latter 22,692 acres. Cabbage, rape, vetches, etc., are also cultivated.

Its grass lands are classed as permanent pasture or fields subject to rotation; of the former there are 79,146 acres, and of the latter 62,291 acres.

While orchards and market gardens occupy no extensive acreage, much attention has been paid to them from very early times; the first pear trees of Newburgh were planted by the monks. The mansions of the landed gentry and the residences of wealthy farmers have splendid gardens, and the well-cared-for and fruitful plots adjoining villas and cottages are a marked feature of the entire county. In the neighbourhood of the larger towns there are numerous nursery and market gardens.

Fife is not heavily wooded. In the time of James IV all the woods in the county were said to have been cut down to build the *Great St Michael*. In the days of Mary Queen of Scots, Fife shared the common barrenness of the Lowlands, a state of matters that called forth laws to encourage the planting of trees and prevent the wanton injury of those standing. Since those times a great change has come over the whole face of the country, much attention having been devoted during the two last centuries to arboriculture. Though there are no great plantations in Fife, small areas of woodland abound, many of its hills and crags are crowned with trees, and in some places on the firths, as at Balmerino and Aberdour, woods come down to the very water's edge. Many of the finest plantations are in the vicinity of mansion houses, which may be the result of an old law that everyone possessing a certain amount of land was required to plant three acres of wood round his dwelling. Also the visit of Dr Johnson to Scotland in 1773 gave a certain impetus to planting, in consequence of his having alleged that he had not seen from Berwick to St Andrews a single tree more than a century old. In Kilconquhar as many as 700 acres are under wood, and Largo, Auchterderran, Falkland, and other parishes are not far from reaching that total. Altogether the area of the Fife woods is 19,471 acres, which places it thirteenth on the list of Scottish counties in this respect. There are some very old trees in the county, such as the well-known yews of Forgan.

The Sands, Aberdour

The Fife farmer gives a due share of attention to the breeding and rearing of animals. In the current year (1910) the county owns 10,783 horses, 45,956 cattle, 120,630 sheep, and 5415 pigs. Apart from farming, a considerable number of the horses belong to local hunting studs and to the light horse yeomanry. There is a special Fife breed of black cattle ; and the polled Angus herd of Naughton, and the Clydesdales of Montrave are well known to breeders. Many of the cattle are imported from Ireland to be fed for the market. The higher reaches of the county are largely given over to sheep-farming.

Formerly every farm had its "doo-cot" or pigeon-house, but the destruction caused to the crops led to their disuse, though the quaint buildings still form a common object of the rural landscape.

15. Industries and Manufactures.

Almost the whole of the linen trade of Scotland is carried on in the counties of Forfar, Fife and Perth. Amongst these Forfar easily holds the first, and Fife the second place. The industry is a very old one in Scotland and appears to have taken root in the east, because the soil there is suited to the growing of flax. The fibre is now, however, imported. The most momentous change in the history of this branch of the textile industry took place when about the beginning of last century the hand-loom gave place to the power-loom. Before that the weaver worked in his own cottage: "the young women worked the linen for their first home, the old for their last resting-place." In the old days the industry was widely distributed over the county, but as it ceased to be a home occupation and the weaver had to work in the factory, weaving gradually came to locate itself in definite districts, and the town increased at the expense of the country.

Newburgh and Tayport, in the north of the county, have long been small centres of this manufacture. In 1793 the greater part of the inhabitants of the north-western burgh were weavers. Tayport has both a jute and linen spinning-mill and two linen factories. Together Newburgh and Tayport employ between 500 and 600 people in spinning and weaving. The Eden district, with nearly 3000 operatives, is more important in the linen trade. It includes Cupar, Auchtermuchty, Strathmiglo, Falkland, Freuchie, Kettle, Ladybank, and some smaller places. The valley of the Leven, containing Leslie, Markinch, Leven, and many small villages, has between 1000 and 2000 spinners and weavers.

In each of these districts and in others bleaching and dyeing, so closely associated with the staple industry of the county, are largely represented. The bleach-fields of Fife are considered so good that manufacturers outside its boundaries send their yarns to be bleached there.

But we have yet to speak of the two greatest centres of the linen manufacture in Fife. There were weavers in Kirkcaldy (along with which Dysart, Pathhead, and Kinghorn may be considered) as early as 1672. Machinery was introduced in 1792, and steam power in 1807. The first power-loom weaving-mill was erected in London in 1812, and it was in Kirkcaldy that the second successful attempt of the kind in the whole country was made in 1821. Over 3000 people in this district are employed in the manufacture of linen.

To think of the linen manufacture in Fife is, however, to think of Dunfermline. Indeed it is the chief seat of the table-linen manufacture in the world, and the annual output of its celebrated looms has a value of more than £1,200,000. In the beginning of the eighteenth century, it produced huckabacks and diapers; but the making of damasks was introduced in 1718. Improvements were effected in 1779 and 1803, and the coming of the Jacquard machine in 1825 established for the town a supremacy in this manufacture that it has never lost. Nearly half of its products go to America. In Dunfermline there are more than 4000 looms constantly at work, with employment for more than 6000 persons.

Kirkcaldy is almost as renowned for floorcloth and linoleum as Dunfermline is for damasks. In 1847 the late Mr Michael Nairn, in spite of the taunt which gave to his establishment the sobriquet of "Nairn's Folly," established this industry. Linen floorcloth was already made there and elsewhere in Fife, but the new invention used more durable material. Mr Nairn employed the fibre of cork and oil-plant, and the product is even yet without a rival. Between 1000 and 2000 men are required in the local manufacture, and they turn out floorcloth of the annual value of half a million pounds.

Amongst Scottish counties Fife is surpassed in the

Dunfermline

manufacture of paper only by Midlothian and Lanark. There are paper-mills in the neighbourhood of Leslie, Markinch, Leven, Inverkeithing, and Guard Bridge. The output of these rose from 5395 tons in 1885 to 9641 tons in 1892. Various kinds are made—cartridge, rolls for newspaper printing, blotting-paper, commercial papers, etc.

There are extensive iron works at Dunfermline and Kirkcaldy, and much machinery is made in the engineering works of the latter town—gearing for mines, marine engines, boilers, and for the colonies, machinery for sugar and rice mills.

The excellent clays of the county are utilised in the making of bricks and tiles at Leven, Falkland, and elsewhere; and there are terra-cotta and pottery works at Dunfermline and Kirkcaldy. These towns, too, have large corn, flour, and meal mills, while Dunfermline and Anstruther have tanneries.

Shipbuilding, at one time a much greater industry in Fife than it is now, is carried on at Kincardine-on-Forth, Inverkeithing, Kinghorn, and Tayport. There are large saw-mills at Tayport, Leven and Auchtermuchty.

Brewing is carried on at Dunfermline, Kirkcaldy, Ladybank, Anstruther, and Falkland; and distilling at Burntisland and Cameron Bridge.

In connection with the fisheries of the county, Dunfermline, Kirkcaldy, Leven, and Anstruther, have establishments for the making of ropes, sail-cloth, oilskins, and nets.

Salt-making was formerly an important industry in Fifeshire. There were salt-pans at West Wemyss, Inverkeithing, Dysart and other places. Dysart was so famous that a Fife proverb "salt to Dysart" equals "coals to Newcastle." It may be added that in Scotland during the seventeenth and eighteenth centuries the salters and the colliers were serfs, bound by law to remain at the salt-works and coal-mines, and to pass with these when transferred to new owners.

16. Mines and Minerals.

Mining and quarrying are certainly amongst the most important industries of Fife, for the county has long been noted both for the variety and abundance of its minerals.

By far the most valuable of the minerals of Fife are its coals. As early as 1291 the proprietor of Pittencrieff, near Dunfermline, granted a charter to the monks of that abbey to work coals on a certain part of his estate. This shows that the use and value of coal must have been known at a still earlier date in this district, and hence Fife has the distinction of being the first district in Scotland in which coal was worked. When King James VI described the shire as "a gray cloth mantle with a golden fringe," he had in mind the fringe of towns and villages. Had he but known, the agriculture, and still more the coal-mines, were destined to make Fife far richer than its small ports. In James's reign coal-mines had been opened along the shores of the Forth; but since his day these have been developed and extended not only far inland but even under the very waters of the Firth. Indeed it will appear from what is about to be said that Fife, though for a long time out-distanced by western rivals, bids fair to be the chief coal-field of Scotland. The Royal Commission, in their report dated 1905, say :—

"The County of Fife, with the smaller quantities in Kinross, takes the leading position in Scotland in the matter of its coal resources. Besides the coal in these counties, probably two-thirds of that under the Firth of Forth will be worked by collieries in Fifeshire, so that the available resources at less than 4000 feet deep will amount to something like 5,700,000,000 tons, or sufficient to maintain the present output for 930 years. While the output from the Firth of Forth will not rapidly increase for a great many years, when so much coal is available under the land, the output from Fifeshire is certain to advance till it occupies a leading position in the Scotch coal trade."

The output per person employed under ground in 1905 in Clackmannan was 416 tons, in Edinburgh 403, Fife 457, East Stirling 422, Ayr 434, Lanark 417, West Stirling 388, and Renfrew 450 tons. In the same year 7,241,439 tons were brought to the surface, and the number of persons employed in and about the mines was 19,607. In 1908 this large total had increased to 8,412,855 tons, a figure short of that of the previous year by more than 100,000 tons.

The following table is an eloquent tribute to the rapid increase of the industry :

Quantities of coal shipped from the Clyde, the Forth and Fifeshire:

		1899. Tons.		1906. Tons.
Clyde	...	3,559,839	...	4,789,898
Forth	...	3,078,589	...	4,147,482
Fife	...	2,355,613	...	4,995,469
		8,994,041	...	13,932,849

This increase has been largely due to the dock accommodation provided within the last 23 years at Burntisland and Methil: of the above shipments from Fife in 1906, Methil is responsible for 2,793,257 tons and Burntisland for 2,013,454 tons. Other but far less important harbours in the coal trade are Dysart, Charlestown, and Wemyss. Most of the coal exported goes to Germany, Denmark, Sweden, France and other European countries, and much yet farther afield. It has been realised by coal companies in "The Kingdom" that the success of their operations in the future must depend largely on the depth of their mines. Thirty-five years ago there was not a pit in Fife deeper than 120 fathoms, but quite recently 300 fathoms have been reached, and in one case, Glencraig, coal is being mined at a depth of 380 fathoms. Rate of output per day has vastly increased also. Forty years ago 100 tons a day was regarded as a maximum; but at Bowhill as many as 2500 can now be brought to the surface. While most of this activity belongs at present to the west and south of Fife, the hope is entertained that farther east in the county seams may be struck that will be richly productive.

The Docks, Methil

Ironstone of various qualities, of which the best are blackband and clayband, is found throughout the Carboniferous system and particularly in the south-western coalfields of Fife. Only the first of these varieties however is worked at present in the county and that to a less extent than formerly, some of the blast-furnaces being either disused or even entirely removed. The clayband, which requires more coal for its calcination, though more abundant, is less profitable and therefore is little worked. It awaits improved methods of production. In connection with the more earthy varieties of ironstone are found certain pigments, specially ochre, which, obtained in the neighbourhood of Leven, used to be ground in mills there and shipped to Leith for sale. But this was so seldom pure and often so difficult to free from sand as to be unprofitable. The industry has been abandoned for about fifty years.

In certain parts of Fife veins of galena are found. Towards the end of the eighteenth century a mine was worked not far from the summit of the East Lomond Hill. A considerable quantity of lead and even a little silver were obtained. The vein runs north-east and south-west through sandstone and limestone.

Valuable seams of fire-clay are found amongst the coal-bearing rocks. Some of these are used only locally for bricks and pipes in connection with the collieries; but at Lochhead, near Dunfermline, Hill of Beath, Buckhaven and elsewhere, fire-bricks and sanitary appliances are manufactured extensively. Brick-clays are found near the 100-feet raised beach and are worked to the south and west of St Andrews, in the valley of the Eden up to and beyond Cupar, and at Crail, Anstruther, Pittenweem, and other places on the south coast.

Shales sufficiently bituminous for the distillation of mineral oil are found at various points in the southern section of the county, where the carboniferous rocks prevail. At Burntisland, Kinglassie and Pitcorthie near Crail, oil-works were until recently in operation; but foreign competition and the consequent fall in the price of oil rendered these unprofitable with the result that they were closed. Crude and refined oil, and sulphate of ammonia were produced, and also "scale" from which candles were made in a factory near Kinghorn. Under more favourable economic conditions the industry might be revived.

Another useful mineral and one widely distributed throughout southern and eastern Fife is limestone. There is considerable variety in the nature of the stone. From some kinds cement is obtained, from others the lime so much associated with agriculture, building, and many industrial processes. The lime-kiln, usually built against the side of a hill, is a familiar object on many a farm-steading, and it was at one time in regular use preparatory to the manuring of the fields. But lime is no longer so abundantly employed as formerly for this purpose, and most kilns are now disused and are fallen into ruin. A remarkably fine-grained and pure limestone is obtained between Burntisland and Aberdour. Plentiful supplies are quarried or mined at Charlestown, Roscobie, Leslie, Kingsbarns, and at various places in the parish of Ceres; but the most extensive operations in the excavation of this material are carried on near Cults.

Fifeshire is especially rich in building-stone. There is an abundance of sandstones both from the Old Red and the Carboniferous systems. In colour these vary: we find different shades of grey, yellow, and red. A specially durable and valuable kind is got in the Grange and other quarries near Burntisland. The sandstones of Fife are not only of good appearance; they are close-grained and firm, yet easily worked; and the best of them, as in the case of the stone of which the Tower of St Regulus at St Andrews is built, will last for centuries in almost perfect condition. Certain kinds of limestone stand the weather well and are otherwise suitable for architectural purposes. Even igneous rocks, such as certain dolerites and basalts, though less easily dressed, are serviceable for building houses. But such material is more commonly used for constructing walls and field-fences; while great blocks of coarse-grained dolerite are quarried for piers and docks.

The county contains abundance of stone suitable for the metalling, causewaying, and kerbing of roads. The andesites of the Ochils and their eastern spurs are excellent for these purposes, and numerous quarries close to road or railway furnish a ready supply. The basalt-dykes and dolerite sills of the central and southern districts make the distribution of this material a wide one, and the practically inexhaustible supply should not merely suffice for local requirements but should be available, from the means of transmission afforded by the railways and docks of the county, for the needs of less favoured districts. The quarries of North Queensferry, which are close to the Forth, are well known for furnishing such material.

The abundance of coal in Fifeshire has led to the disuse of peat as fuel; and yet for centuries peat was used in Fife as throughout the rest of the country. It is still dug in Beath and Kennoway for the purposes of distillers in flavouring whiskey.

At Inchrye, near Newburgh, there occurs a layer of shell-marl twenty acres in area and two or three feet in depth, which was at one time sought after as a manure. It was probably formed here, as in some other places, under the bed of a lake that has disappeared.

17. Fisheries and Fishing Stations.

In an island country like Britain, surrounded by shallow seas that abound with fish, one of the chief industries is naturally fishing. Not only do our own people use vast quantities of this cheap and wholesome food, but we cater for millions in such countries as Germany and Russia, and send yearly tens of thousands of barrels even as far as Siberia and Central Asia. The importance of Scottish fisheries in particular may be gathered from some of the returns for the year 1908. The Scottish fishing fleet numbered 10,078 vessels, and 8,645,252 cwts. of fish were caught, representing a value of £2,512,162. And this, as reports from almost every district show, was yet one of the poorest years on record. In 1907 the catch was greater by 372,901 cwts., and the value by £636,965.

Fish are caught off all our Scottish coasts, but the east is more important than the west, partly because the North Sea is one of the finest fishing grounds in the world, and partly owing to the remoteness from markets of the north-western and Hebridean fishing ports.

The waters fished by British vessels contain no fewer than 50 species of marketable value. By far the most important of these is the herring, "the poor man's salmon"; and the others are usually classified as round fish, flat fish, and crustaceans or shell fish. Some are caught in estuaries, others off shore, but most, far from land on the deep sea grounds. It is found that some fish, like herrings and mackerels, live in shoals near the surface of the sea, but that others, of which turbots, haddocks, soles, and plaice are examples, feed deep down, often as many as 90 fathoms, below sea-level. The former, or *pelagic* species, are caught mainly by drift nets; the latter, or *demersal* kinds, by lines or by nets trawled along the sea bottom. Near the land nets are often attached to stakes driven into the ground, the so-called fixed nets. Fife fishermen prosecute their calling in the two great estuaries that bound the county and along the ocean foreshore; while those who go to a greater distance fish between Fife Ness and the Bell Rock, or as far as 200 miles east and east-south-east of the May Island.

One associates with fishermen the huge boats, well tarred and quaintly lettered, that are so often seen at seaside places drawn up on the beach; or to the mind's eye comes such a picture as Tennyson describes (he might be speaking of some Fifeshire fishing village) where

> "Long lines of cliff breaking have left a chasm;
> And in the chasm are foam and yellow sands;
> Beyond, red roofs about a narrow wharf
> In cluster."

The sun is setting and the boats are leaving the harbour to the cheery "yo-ho!" of men and boys, as they ply oar and rope. Up go the brown sails and the boats glide out before the evening breeze into the clear blue waters of the bay. But such scenes will soon belong to the romantic past, for sails and oars are being rapidly superseded in the fisherman's craft by steam and motor power. It is found that smaller boats are somewhat inefficient, and that if Scottish fishermen are to compete successfully with rivals at home and abroad, they must have larger, more effective, and better equipped vessels. Steam has enabled fishermen to carry on their work in calms and even during rough weather. On many of the larger boats curing is done while the vessel is at sea; and fresh fish can be taken direct to distant markets. The rapidity and extent of the change may be gathered from the fact that in 1898 the Scottish fishing fleet consisted of 11,576 boats with a value of £2,029,384; while in 1908, though there were only 10,078 vessels, their value had risen to £5,223,149. Nevertheless 87 per cent. are still sailing vessels; and Fifeshire is more conservative in this respect than many places, nearly 90 per cent. of its fishing craft being of the older type.

Fifeshire is naturally a fishing county. An industry

that gives employment, directly or indirectly, to more than 3000 people and produces about £70,000 in a year (1907) is not lacking in importance. From Kincardine-on-Forth to Newburgh-on-Tay there are no fewer than 25 fishing stations. These are controlled from two centres: Leith for the Fifeshire stations on the Forth

Pittenweem and Harbour Head

west of, and including, Wemyss; and Anstruther for the remainder round to Newburgh. The estuarine fisheries of the upper Forth and Tay are those of sprats and sparlings; mussels are taken in large quantities at the mouth of the Eden and at Tayport; herrings, codlings, haddocks, plaice, and dabs in the wider part of the Forth and in the North Sea generally, while lobsters and crabs are captured in large numbers near Anstruther and more particularly off Crail. Herrings are by far the most important catch. Taking the returns from the Anstruther district alone, we find that 95,342 cwts. of these fish were taken in 1908. Next came cod (13,556 cwts.) and haddock (5594 cwts.). Plaice and dabs were next in order. The total catch was 117,734 cwts. valued at £33,602.

These were taken chiefly by trawlers; and the returns show that steam vessels, whether drifters or trawlers, far outdistance in utility the older kinds of vessels, and that the net is fast superseding the line as a means of capture. The area of Anstruther nets, second only in Scotland to that of the Shetland district, is 21,120,028 square yards.

Fish are used either in a fresh state or are cured and packed in barrels for export. In the former case, quick transit is absolutely necessary, and here the railway is an invaluable auxiliary to the fisherman. In 1908 the returns of fresh fish for the Anstruther district were 62,797 cwts.; and most of these were carried by express train from the southern Fife ports to Dundee, Glasgow, Edinburgh, and the English markets. By means of ice and cold storage, swift steamers can now convey fish in good condition from the fishing grounds themselves direct to distant British and even continental ports.

Occupations ancillary to the fisheries are chiefly those in connection with the curing and packing of herrings and certain kinds of white fish. In the season not only is the population considerably augmented by fisher folk from a distance, but there is a busy time for a whole army of curers, coopers, gutters, packers, clerks, labourers, carters, and hawkers. In 1908 the Anstruther fishery officer had a total to report of 8553 barrels for the home and export trade.

18. Shipping and Trade.

In the course of our peregrination round the shores of Fife, reference was made to the numerous little ports of the county. Their exports vary with the industries of their localities, and their imports with the special requirements of the several communities. Thus agricultural districts ship grain and potatoes, and import lime and guano. Manufacturing centres, again, send out linen goods and receive timber for saw mills and esparto grass for paper mills. Coal is imported where there are no local mines.

Certain circumstances have contributed to the decay of several ports and retarded the growth of others. Prominent amongst these is the development of railways. Thus Kincardine-on-Forth with its excellent ferry at one time controlled the cross-country traffic of the upper part of the estuary, but the main through lines from north to south now lie far to the west or the east and have left it isolated. The Tay Bridge has deprived Tayport, and the Forth Bridge Burntisland, of most of its ferry traffic. Such industries as distilling, brewing, coal-mining, and ship-building have become exhausted or even extinct at Kincardine-on-Forth and other places. Shallow harbours, like that of Inverkeithing, prove inadequate for the requirements of modern vessels; while others, like those of St Andrews and Leven, are very difficult to enter. In the latter case thousands of pounds have been spent in a vain endeavour to break through a barrier of sand impassable for large vessels at the mouth of the harbour.

The three chief ports are Kirkcaldy, Burntisland, and Methil.

As a port Kirkcaldy used technically to include the various "creeks" from Fife Ness to Downing Point, but Burntisland and Methil are now regarded separately. The history of Kirkcaldy as a port is sufficiently chequered. In the old days it traded with various parts of Great Britain and the Continent in coals, salt, and salted fish. About the middle of the seventeenth century it owned 100 vessels; but soon afterwards 94 of these (of the value of £53,791) were lost, and the Dutch wars after the Restoration so crippled its trade that in 1687 its customs returns fell to one half of their former value. Thereafter trade revived, but only to be again checked by the Union. In 1768, 1788, 1831, and 1883 it owned vessels severally as follows: 11, 30, 95, and 18. Previous to 1843 its harbour was only tidal; but great improvements have been effected since then. Its present importance may be estimated from the latest returns of its exports and imports. In 1908 it exported coal, coke, and other articles to the value of £48,288; and imported cork, flax, hemp, oilseed, paper-making materials, seeds, stones, wood, and other goods, to a total of £21,328. In the same year the tonnage of outgoing vessels was 56,143, and that of vessels entering, 29,024.

The Docks, Burntisland

Burntisland was at one time the principal rendezvous of the fishing boats of the Forth and East Coast fisheries, and its trade in herrings was particularly vigorous. But its importance in these respects fell off. Under the fostering care of the North British Railway Company, on whose system, prior to the building of the Forth Bridge, it was an indispensable link, immense improvements were made at the harbour, no less than £150,000 being spent in the construction of a wet dock, a sea wall, and other works. Much more than this has been done since, with the result that the harbour revenue grew from £197 in 1860 to £14,785 in 1880. In 1908 it imported goods to the value of £118,314, and exported articles, chiefly coal, to that of £1,113,655. The tonnage of vessels that entered Burntisland with cargoes is returned for that year as 604,613; while the tonnage of those that cleared the port was 843,510.

The growth of traffic at Methil dock is phenomenal. The dock was opened in 1887, when 219,884 tons of coal were shipped. Ten years later the output of coal amounted to 1,090,324 tons; and it rose in 1904 to 1,985,826 tons. The inner dock has an area of 4¾ acres, and the outer dock, which was opened in 1897, is 6½ acres in extent. The imports of Methil for 1907, which comprised flax, hemp, oil-seed cake, paper-making materials, wood, etc., had a value of £331,069; and its exports, chiefly coal, of £1,352,393. It ceased to be ranged under Kirkcaldy in 1904, but the above returns include those of Anstruther, Largo, and Leven.

19. History of the County.

Fife has played a conspicuous part in the history of Scotland. Separated from the Anglian Lothians by the wide estuary of the Forth, it lay for many centuries in the Pictish district of Scotland, and formed one of the numerous independent kingdoms of that obscure folk. Those dim Pictish times are known to us in Fife, if at all, mainly through the early half-legendary labours of St Serf and his fellow Culdees. A cave at Dysart, which may have been his cell, an island in Loch Leven, a well at Alva, and a bridge in the Ochils, indicate the district over which his influence prevailed, and Culross is said to hold his tomb. The site of a Culdee monastery—Kirk Heugh—is still pointed out in St Andrews, but the city is that of St Rule, whose square Byzantine tower is one of its most conspicuous and venerable monuments. This saint is said to have brought from Patras the relics of St Andrew, and the tradition may at least serve to remind us of the fact that Culdee was superseded by Roman Christianity, a change, however, that was not fully carried out till some centuries later.

With the dawn of more trustworthy history in the reign of Malcolm Canmore (1057–1093), the main interest in Scottish annals finds a centre in Dunfermline. His good queen, Margaret, landed at Queensferry in 1068, and from Dunfermline, her husband's capital, there emanated during the years of their beneficent reign an influence that left a deep impress on the whole country. Her vigils, her prayers, her alms, her good deeds, made her the latest of the saints of the Scottish Church.

The reign of Malcolm and Margaret marks the end of the Gaelic regime in Fife and the beginning of Norman influence in Scotland. Families of French ancestry gradually settled in the county under the Anglo-Norman kings; the Feudal System here as elsewhere was adopted, and Culdee faded before Roman Christianity. In course of time certain of the kings—David I and his successors—granted charters to towns. St Andrews, Inverkeithing, Dunfermline, Kinghorn, Crail, and Cupar, were amongst the first to be thus favoured; and to these monarchs must be ascribed the establishment of many ecclesiastical foundations throughout the county. The period of Scottish history prior to the War of Independence closes in 1286 with the tragical death of Alexander III.

Wallace gained a victory at Black Ironside, or Earnside, near Newburgh, and perhaps drove the English out of Fife; and in the flush of victory Edward I visited Dunfermline for the purpose of receiving the homage of the local nobles. On the 5th of July, 1318, Robert the Bruce was present at St Andrews on the occasion of the dedication of the cathedral; and he was buried at Dunfermline, where in 1818 his skeleton was unearthed.

Fifeshire occupies a prominent place in the general history of Scotland at the dawn of the Reformation. St Andrews attained a bad eminence by the judicial murders of Patrick Hamilton (1528) and George Wishart (1546). Retribution came swift-footed. On the 28th of May in the same year Norman Leslie and his accomplices forced their way into St Andrews Castle and murdered Cardinal Beaton. The preaching of John Knox at St Andrews inaugurated stirring times. His great voice thundered anathemas from the pulpit of the parish church against the monuments of idolatry which he saw around him, and his vehemence is graphically described by a con-

John Knox

temporary—"He was like to ding the pulpit to blads and fly out of it." It had an immediate effect. Not only was the venerable cathedral destroyed, but a civil war was entered upon by the Lords of the Congregation, which, after the meeting of the rival forces at Cupar Muir, drove the Queen-mother and her French supporters from Fife, and resulted in the establishment by the Scottish Parliament of the Protestant religion in 1560.

In 1637 the Scottish Covenanters found an able and dauntless leader in Alexander Henderson, minister of Leuchars. He was indeed the author of the National Covenant and moderator of the Glasgow Assembly that insisted on the deposition of bishops and the establishment of presbyterian church government. The Leslies, of Duns and Philiphaugh fame, were both Fife men. Samuel Rutherford, afterwards Principal of St Mary's College, St Andrews, and George Gillespie, were closely associated with Henderson in London, when as commissioners from the Scottish Assembly they took a leading part at Westminster in 1643 in drawing up the Solemn League and Covenant, the Confession of Faith, and the Catechisms. The regime of Cromwell in Fife, which was inaugurated in 1651 by the victory of his forces at Pitreavie near Inverkeithing—the last battle fought in the county—was of immediate and lasting benefit; for Cromwell and Monk freed the middle classes for a time from the tyranny alike of the nobles and the presbyterian clergy. The Restoration, of course, temporarily undid much that had been thus accomplished; but prosperity was in the long run assured. The name of Archbishop Sharp links Fife to general Scottish history under the restored Stuarts. His murder by Balfour of Kinloch (Scott's Balfour of Burleigh), Hackston of Rathillet, and others on Magus Muir near St Andrews, caused one who had made himself notorious to friends as a self-seeker and to enemies as a persecutor, to be regarded as a martyr for the episcopal faith.

Deprived by the Union of Parliaments of much opportunity for political discussion, Scotland after 1707, and

Archbishop Sharp's Monument, Strathkinnes

especially Fifeshire, plunged with the greater zest into theological controversy. "The region between the Tay and the Forth," it was said, "is the hottest quarter of religious zeal and controversy in Scotland." The result was schism within the presbyterian church, and the hiving-off from its main body of many sects. The rocks of offence were state control of the church and appointment of ministers by patronage. Western Fife was the home of the Original Seceders, and at Port Moak, near Kinross, in 1733, thousands would assemble on a Sabbath to listen to the preaching of Ebenezer Erskine. With more than the proverbial caution of the Scot, Fifeshire men eschewed intermeddling in the Jacobite rebellions, and outside of religious movements little falls to be recorded of Fife in the eighteenth and nineteenth centuries.

20. Antiquities—Prehistoric, Roman, Celtic.

For any knowledge we have of the men of a district in prehistoric times, we are indebted to the research of archaeologists. Many relics of primitive man and his ways have been dredged in rivers and lakes, or unearthed in fields and mounds and hill-tops. The archaeologist examines these, compares them with similar remains found elsewhere, and is thus able to deduce many reliable inferences as to the aborigines of the district in question.

Now Fife is rich in such relics: they have been found in places widely distributed throughout its boundaries, and are believed to represent several distinct primitive periods. It is usual to divide those early times into the Stone Age, the Bronze Age, and the Iron Age. In England, though not in Scotland, the Stone Age divides into, first, the palaeolithic period, when tools and weapons were roughly shaped or chipped; and, second, the neolithic period, when tools and weapons were finely chipped or polished. Only the neolithic type of implement is found in Scotland.

About the middle of last century there was found in the bed of the Tay at Ballinbreich, near Newburgh, an interesting relic of the Stone Age. This was a granite stone axe, wedge-shaped and perforated, which measured 5½ by 2 inches. Not far off, near Mugdrum, a whinstone axe was discovered; a bronze sword 30½ inches in length at Newburgh; bronze daggers at Kilrie and Auchtermuchty; a bronze axe at Kingskettle; and a bronze armlet at Kinghorn.

Two of the oldest relics of the primitive natives of Fife are a couple of boats found some seventy or eighty years ago near Newburgh in the bed of the river. They were hollowed out of single trees and the larger is 28 feet long.

Indications of very early agriculture have been traced in some places at a height far above what modern farmers would deem suitable for ploughing. But in those days high ground was at least free from marshes and would probably have the advantage of being open and easily defended. Indeed it may be inferred that in times of peace, and certainly in times of war, the early natives of Fife lived amongst the hills. Hill forts are more numerous than remains of lowland dwellings. On Norman's Law there is one of these strongholds. Circular earth-works of considerable extent were thrown up round a space

Palaeolithic Flint Implement
(From Kent's Cavern, Torquay)

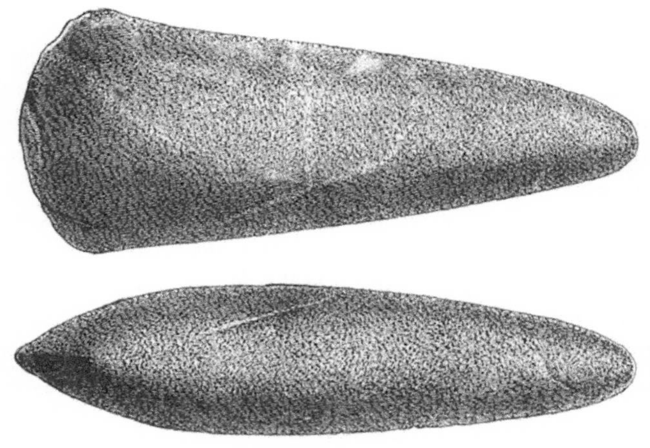

Neolithic Celt of Greenstone
(From Bridlington, Yorks.)

large enough to enclose not only the forts in which the people would take refuge but also the pasture on which their flocks would feed. A still more interesting example is that on Craig of Clachard in the parish of Abdie. The fort in this case is not only more extensive than that on Norman's Law, but it gives evidence of far more labour and constructive skill.

More curious still are the underground earth-houses of olden times. Two specimens have been found in Fife—one at Pirnie, near Wemyss, and the other at Elie. In each case the whole structure is subterranean, and consists of a low entrance under the level of the surrounding ground. The dwelling can be entered only on hands and knees, but as one advances it widens and increases in height. The entrance is strengthened with heavy lintels; and a peculiarity of these Fifeshire earth-houses as compared with others elsewhere is that steps had been cut to lead down to them.

Remains of lake-dwellings, or crannogs, similar to those found in other parts of the British Islands and on the Continent, have been discovered in the Fife peninsula. These were formed on piles driven into marshy ground or on an artificial islet in lakes at some little distance from the shore, which was reached by a causeway or gangway that could easily be destroyed in case of attack. Two specimens of the marsh crannog in Fife are those of Collessie and Stravithy. They are constructed mainly of stone, not, as was usual, of logs.

Tumuli, or burial-mounds, form another interesting class of prehistoric remains. From these evidence is obtained of the customs, the skill, and even the creeds of the aborigines. In 1876 one of these cairns was opened near Collessie. It consisted of a mass of stones and was 120 feet in diameter and 14 feet high. Near the centre was a cist that contained a human skeleton and a bronze blade mounted with gold. Decorated urns were also found. Similar remains have from time to time been discovered at Dairsie, Kingsbarns, Pitreavie, Tayfield near Newport, and many other places. The Tayfield urns are specially interesting: in one case nine were found placed around a central urn in a circle 14 feet in diameter. The decorative ornaments and the weapons found showed that those buried there were men of the Bronze Age, and there was a necklet of beads and plates of jet in one of the cists. That weapons and ornaments should have been interred along with the dead seems a sure indication of the belief in a life beyond the grave; and yet the presence of such objects in a tomb seems to point to pre-Christian times and to be distinctive of heathen burials.

Norrie's Law, situated on high ground about three miles to the north of Largo Bay, is one of the most important and interesting of the burial-mounds of Fife. It is a pagan structure with a circumference of 160 feet. In the sand were found silver ornaments, and in 1819 a pedlar surreptitiously abstracted and sold a quantity of these, the weight of which was afterwards estimated to have been 400 oz. It is thought, however, that the ornaments had been secreted in the mound at a date much later than that of its origin.

The standing stones of Lundin are said to be burial stones of Danish chiefs; and human skeletons which at various periods have been found near the south coast of the county are regarded as evidences of the conflict with the invading Danes in the ninth and following centuries.

At Scoonie, Largo, Lindores, and Crail have been found sculptured stones of great archaeological interest.

Standing Stones of Lundin

With the exception of the Lindores Stone they are unmistakably the work of early Christian times. The Scoonie Stone is thus described in the *Proceedings of the Antiquarian Society of Edinburgh*, in whose museum it is preserved:—

"The stone measures 3 feet 6 inches in length, 2 feet

4 inches in breadth, and 4 inches in thickness. It displays on the upper part the so-called "elephant" or beaked animal, its extremities terminating in scrolls; and below it apparently a deer hunt; a rider on horse-back, and in front of him a dog on the point of seizing a full antlered stag. Below these is another horseman, and in front of him a dog, and below the dog a third horseman. At the lower angle of the stone is cut a small cross. Along the whole left of the stone is incised an Ogham inscription. On the reverse is sculptured a Latin cross, with a plain circular disc in the centre, and the limbs filled up with interlaced rope or knot work; the scroll termination of the nondescript animal (like the dog-headed animals on the Ulbster and Brodick Stones) appears over the left limb of the cross."

Like the crosses of Iona, though these are more highly decorated and probably much later, the stones of Fife must belong to a time subsequent to that in which Columba taught. The Stone of Crail has a rude representation of the Virgin and Child carved on it. The Lindores Stone has no indication on it of the Christian religion. Besides these, other stones have been found at Abercrombie, Dunino, Inchcolm and elsewhere.

Stones bearing curious markings—cuppings arranged in circles and rectangles—that are a puzzle to antiquarians have been found at Torrie and Pitcorthie in the parish of Carnbee. They are thought to be at least as old as the Bronze Age. To early Culdee missionaries and hermits is ascribed the curious carving found on the walls of many of the caves along the shores of Fife. Caiplie Cave, in the parish of Kilrenny, the traditional retreat of St Adrian, is one of the most interesting of these; and the cave of Kinkell, near St Andrews, has numerous crosses incised on its walls. Jonathan's Cave and the Doo Cave at East Wemyss have the most elaborate and important markings.

Mugdrum Cross, near Newburgh, is one of the most curious and interesting of the antiquities of Fife. The shaft is 10 feet 10 inches in height, and may have been a regular cross, though the arms have disappeared. The ornamentation on it is Celtic in character, and there is a pictorial representation of a boar-hunt, or what is supposed to be such. It is probably as old as the time of St Columba, and may be symbolical of the passing away of the older heathen worship and its replacement by Christianity.

Macduff Cross, near Newburgh, is the centre of an interesting tradition. It is said to have been erected by the Thane of Fife after his escape from the vengeance of Macbeth. The cross itself is indeed destroyed, having perhaps been broken by the Reformers in 1559, but the pedestal remains, with nine curious markings, really of natural origin, but according to legend the socket-holes of rings at one time attached to it.

It was a place of sanctuary for any related even in the ninth degree to Macduff, who, having without premeditation committed manslaughter, sought refuge near it. Special rites had to be observed by the refugee and an offering of nine cows must be made in order to escape punishment. If these observances were not strictly followed out, or if the necessary consanguinity to

Mugdrum Cross, Newburgh

Macduff were not proved, the claimant was there and then put to death and was buried on the spot. Hence for many centuries the place was believed to be haunted, a superstition which has been dissipated only within the last few generations.

Cultivation has all but obliterated any traces there may have been of the Romans in Fife. The old historian Sibbald avers that vestiges of their presence were at one time visible near Burntisland; and it has been asserted that Agricola landed there and marched in a north-westerly direction, whilst his fleet circumnavigated Fife Ness and sailed up the Tay. But this is largely conjecture. There seems, however, to have been a permanent encampment at Lochore, traces of which remained at Camp Farm until about 100 years ago. At various places in the county, sword blades, spear heads, and hoards of Roman coins have been found, but while the former points to the certainty of local warfare, the coins (such as the 700 denarii found in the moss at Port Moak) may well have been deposits independent of any actual occupation by the Romans. Sheriff Mackay quotes with approval the lines of an old Fife poet who thus addresses his native county:

> "But thou didst scorn Rome's captive for to be,
> And kept thyself from Roman legions free."

21. Architecture—(*a*) Ecclesiastical.

The styles of architecture. for several centuries after the year 1000, are distinguished from one another mainly by the form of the arch. The Norman was the earliest of these and it is characterised by its rounded or semi-circular shape. In Britain this style was in vogue from the Conquest to the twelfth century, when it was succeeded by the Pointed or Gothic style. This lasted with variations until the fifteenth century. In the thirteenth we have what is known as the Early English or First Pointed period; then, after a transition, we find in the fourteenth century the Decorated period, distinguished by its window tracery. A third Pointed Period is known as the Perpendicular. Its name describes it—it was a time when loftier, stiffer lines were favoured, and when the arches were more flattened than in the earlier periods. This remained until the Renaissance. In Gothic, or Pointed architecture, stability was attained through the balanced counterpoise of a vast series of pressures—oblique, perpendicular, or horizontal; but in Greek architecture there was no pressure that was not met by vertical support, and hence we have column, entablature, frieze, and cornice, as essential features. Rapid as was the change from the one style to the other, it was not abrupt. There was again a transition period during the sixteenth century, which as it advanced displayed Tudor and Elizabethan peculiarities, often involving a mixture of the Gothic with the Renaissance characteristics. In addition to these outstanding styles, we meet with varieties—Italian, Spanish, French, Flemish, German—in which the special genius of each nationality displays itself.

Not only do buildings erected at various periods differ in style, but very often we find such differences in one building, the reason being that the construction of a cathedral, for instance, would not be completed in a single century, but by the architects of successive generations, whose knowledge and taste advanced with their times. The material, too, varied with age and country—from the marbles of Greece and Italy to the chalks and flints of south-eastern England, and the freestones of Scotland.

Ecclesiastical buildings have different designations according to the purpose for which they were originally intended—cathedrals, churches, abbeys, monasteries, priories, and the rest: but yet the name is not, at least now, always a guide to their use—an abbey is often only a cathedral with another name; a priory may be a mansion. Few districts of similar area are richer in monastic foundations than Fife, the piety of Canons Regular, Dominicans, Benedictines, Cistercians and Franciscans being represented in the county by buildings now mostly in ruins.

A typical example of primitive architecture is the oratory of the hermit of Inchcolm. Alexander I, as a thanksgiving for preservation from shipwreck, founded a monastery here in 1123. The central square tower still standing was probably a part of the original monastery. The other buildings that form part of it belong to different periods. In this monastery, which is probably as old as that of Iona, is specially to be noted the stone-roofed octagonal chapter house, one of the finest in the country.

Dunfermline Abbey is one of the most interesting and important ecclesiastical buildings in Scotland. Architecturally it combines Anglo-Norman and Gothic features. The Abbey church succeeded Iona as a place of royal

The Monastery, Inchcolm

sepulture, and the new parish church, with which it is structurally connected, forms a kind of national mausoleum to King Robert the Bruce, whose remains are there interred. The older parts are Norman, and fine specimens of that style, while the Transitional, the Decorated, and the Perpendicular, are nobly exemplified in later parts of the edifice.

St Andrews Cathedral, destroyed by the "rascal mul-

Tomb of Bruce, Dunfermline Abbey

titude" in 1559 is, or rather was, undoubtedly the finest ecclesiastical building in Fifeshire; and even yet it is majestic in its ruins. Begun in 1159 and completed in 1318, its building was carried on by no fewer than eleven successive bishops. The architecture is partly Norman and partly Early English. The church of St Regulus, within the cathedral precincts and indeed close to the main building, is a square tower 112 feet high. It is probably a much older structure than the cathedral, and may be of Celtic origin. The Augustinian Priory, founded in 1144, was one of the finest of its kind in existence, but almost nothing of it is now left. The stately Gothic Pends, which formed the Priory Gateway, still remain. A picturesque memorial of old St Andrews is the ruins of the north transept of the Blackfriars Monastery, founded in 1274. Amongst the other churches of St Andrews the Town Church and the College Church deserve special mention. The former, the Church of the Holy Trinity, is believed by some to have been originally built in 1112. It is partly Norman and partly First Pointed in style, and contains a monument to Archbishop Sharp. The square tower of the College Church (1456) with its octagonal spire rises to a height of 156 feet, and is one of the landmarks of the district. In the church proper, which adjoins it on the east, there is a monument to Bishop Kennedy.

The tower of the Parish Church of Markinch is one of the purest specimens of Norman architecture in Scotland; there are only four others of which this can be said, one of them being St Andrews. The chancel and

The Trinity or Town Church, St Andrews, where Knox preached in 1547, 1559 and 1571–2, as it was in the eighteenth century

Norman Church at Leuchars

nave of the chapel of Aberdour are also Norman; but the most beautiful and quaint edifice of this style in Scotland is the apse of Leuchars Church. It belongs to the thirteenth century. Specially noteworthy are the rich mouldings of the arch at the entrance of the apse. The fragment is not used as part of the present church.

Ruins of Lindores Abbey, Newburgh

The church of St Monans is an interesting example of the Second Pointed style. Like so many more churches of ancient date in the county, it was at one time partly destroyed. A part of the original building is a stunted tower with a low octagonal spire which forms a conspicuous object in the little burgh and helps to lend it an old-world look.

Lindores Abbey must, to judge by its ruins, have been a building of great architectural beauty. The most perfect portion remaining is the groined arch of its porch. It is a specimen of the First Pointed Period. The Second Pointed style is exemplified in the Pre-Reformation church of Kirkcaldy.

Many of the modern churches of the larger Fife towns are handsome and costly structures and furnish examples of the various periods of the Gothic and the Scottish Renaissance styles of architecture. St Margaret's Memorial Church, Dunfermline, erected recently by the Roman Catholics of Scotland, may be mentioned as an instance: the style is Transitional of the twelfth century.

22. Architecture—(*b*) Military.

The sterner nature of the purpose for which they were built is reflected in the architecture of the castles of Fife, as of Scotland generally. There is no occasion or room here, as in churches, for ornament or decoration. Built on rocky sites amid mountain fastnesses, on the edge of some beetling sea-cliff, or on some islet washed by the waters of loch or ocean, they stand solid, four-square, weather-beaten—memorials of warlike times and warlike men. They are usually plain, gloomy keeps with stepped gables, loop-holes instead of windows, and angular turrets. The oldest of them are in ruins, and such as were built in the sixteenth and later centuries preserve many of the characteristics of earlier times. They would even then be half dwellings, half fortresses. Some of the distinctive features of this Scottish baronial style, for which we are in part indebted to our old allies, the French, are turrets projecting from the wall upon bold corbellings and terminating in pointed roofs; towers of circular plan; parapets and battlements; roofs of steep pitch; gables of stepped outline; small, square windows; plain unadorned door-ways; prominent and lofty chimneys. Here, as in other kinds of buildings, later additions often display a different style of architecture from that of the original plan, suited to the tastes and requirements of a later age; and hence not infrequently the whole presents a somewhat incongruous mixture of styles.

Girt as it is by sea, firth, and mountain range, Fifeshire is itself naturally fortified. For this reason, perhaps, it has had on the whole a more peaceful history than many another county, and for this reason too it has few castles, if any, that may be regarded as fortresses of the first class. Nevertheless the number and variety of lesser strongholds are surprising, and many of them recall feudal times.

Rosyth, a square, thick-walled tower resembling a Norman Keep stands in ruins on the shore of the Forth a little above the Bridge. It is difficult to say when it was built; but from the initials M. R., which are to be seen over its main entrance and the date 1561, it is to be associated with the return of Queen Mary from France. Inchgarvie Island, now the base of the central support of the Forth Bridge, was fortified by command of James IV. Its castle was subsequently used as a prison and a place of quarantine. It is a specimen of an island fortress. The

Rosyth Castle and Forth Bridge

ancient castle of Aberdour is believed to have been erected by Randolph, Earl of Moray, the companion of the Bruce; and, if this is so, must date from the early part of the fourteenth century. Two miles south-west of Kirkcaldy stands the ruined tower of Balwearie, the home of the wizard, Michael Scott, in the thirteenth century. Near Dysart on a commanding cliff overlooking the estuary, is

Ravenscraig Castle

the castle of Ravenscraig, designed, it appears, by James II, as a more reliable stronghold than Falkland in time of war, but not completed until after his death. This is the "Castle Ravensheuch" of Scott's Rosabelle. Farther east, between Wemyss and Buckhaven, is the ruin named Macduff's Castle but of a date long subsequent to that of Shakespeare's hero.

In the hilly region to the north-east of Dunfermline stand the ruins of Lochore Castle, which as the name implies marks the site of a former loch from amidst whose waters rose the fortress. It was approached from the east by a causeway, which is said to have been in existence not so very long ago; and the whole may therefore be regarded as a specimen of a lacustrine stronghold. The

Macduff's Castle, near Leven, Fife

style of the architecture is that of the fifteenth, though report says that there was a castle on this site as early as the twelfth century. On the banks of the Leven in the Parish of Markinch rises the massive and picturesque tower of Balgonie Castle, a possession of the Earls of Leven from the time of Charles I. The building includes a donjon and keep and is interesting as the place where Alexander Leslie, the covenanting general, died in 1661.

Balcomie Castle, near Crail, must have been far more extensive than the remaining portion would indicate. Much even of that is comparatively modern and built in a utilitarian spirit. The old tower, Scottish baronial in

Scotstarvit Tower, Cupar

character, and an arched gateway with the inscription 1602, are now the oldest parts of the building; but as this was the castle that housed Mary of Guise on her arrival from France to marry James V, the part in which she was received must have been far older.

In the parish of Ceres is Scotstarvit Tower, once the

residence of Sir John Scot of Scotstarvit, well known at home and abroad in the seventeenth century as a patron of literature. The architecture is severely plain; but, apart from the above interest attaching to it, it is a good example of the L-shaped style of building and of the union of mansion and fortress.

Dairsie Castle, also a ruin, was evidently intended more as a baronial stronghold than as a dwelling. It seems to date, in part from the fourteenth and in part from the sixteenth century, and may be regarded as an example of French influence in Scottish castle building, and in its latest portions, of the decorations that were added to certain strongholds when more peaceable times dawned.

Episcopal palace as it mainly was, St Andrews Castle is perhaps more connected in comparatively recent historical times with warfare than any other stronghold in Fife. It is situated on a lofty terrace washed on the east and the north by the sea, and must have been in its palmy days a place of great strength. Between the present roadway and the wall of the building is a moat, crossed by a modern bridge that leads to what is now the entrance, though the original doorway was towards the east. On either side are vaulted guardrooms beyond which lies the courtyard with the castle well. In the north-west corner steps lead down to a bottle-shaped dungeon hewn out of the rock, the drear, unwholesome prison of many a luckless culprit, who had to be lowered into it by a rope. The kitchen, the hall, the chapel and the living apartments surrounded the central court, 150 feet square. The walls are in parts twelve feet thick. Built in 1200, the castle witnessed many a stormy scene in Scottish history from the time of Wallace and Edward I onwards, each of whom is said to have occupied it. After the murder of Cardinal Beaton it was frequently attacked, and by 1685 had fallen into a totally ruinous condition.

Creich Castle, in the northern parish of the same name, must, to judge by its ruined tower, have been a place of considerable strength. On one side it was protected by a morass, on the others by outworks. It may be regarded as a good specimen of an inland stronghold. But undoubtedly the castle of Ballinbreich near Newburgh was the strongest fortress in the north of Fife. It faced the river and was protected on right and left by deep ravines, while to the south there may have been a ditch. Its courtyard was divided into an outer and an inner bailey. There was a keep adjoining the south wall, a tower of defence on the side next the river, and a round tower with shot holes in the south-east angle of the walls. Extensive alterations and additions seem to have been made from time to time; but the thickness of the walls, the size of the court-yard, and other architectural details have led experts to assign to the original building a date within the fourteenth century. It was long a stronghold of the Earls of Rothes.

23. Architecture—(*c*) Municipal.

With no fewer than sixteen royal burghs, Fife might be expected to possess many fine municipal buildings; but it must be borne in mind that most of these towns, though places of no little consideration when they received the recognition of our sovereigns, have failed for various reasons to maintain their early importance. Hence in many the town halls and other public buildings are architecturally quaint and interesting rather than handsome. In none do we find specimens of the builder's art that can for a moment compare in beauty of design or richness of decoration with the municipal edifices of certain continental cities. In some cases, indeed, the old buildings have been entirely removed and their places have been taken by modern structures in which utility rather than taste has been observed.

Crail and Dysart may be selected as affording characteristic examples of old municipal architecture. In the former the town-hall is the most interesting public building: its ancient square tower and spire have something of a Flemish aspect. In the sixteenth century Dysart was a place of very considerable commercial importance, and its square, in which stood the town-house, was thronged with merchants and presented a sight often commented on by early writers. The old Tolbooth having been demolished by fire in 1578, another was erected in 1617; but this, too, met with a serious accident. Cromwell's soldiers used it as a magazine and it was almost destroyed by an explosion. It was repaired in a substantial manner and remains to this day, a plain rubble-work structure, with a tower and spire.

Early civic architecture in the ancient burghs is perhaps best exemplified in the irregular but picturesque lines of their streets. The artist in search of old-world "bits" could

Old Street in Kinghorn

not do better than visit Dysart, Crail, or St Andrews, any one of the main thoroughfares of which, to say nothing of the quaint, narrow, crooked alleys that flank them, would form a striking illustration of town life and its external conditions in the middle ages. In most cases these streets were entered through a gate or port, specimens of which yet remain, though to modern traffic they are a source of some inconvenience. The West Port, St Andrews is a good example. It is one of three archways that stood at the ends of South Street, Market Street, and North Street, the main arteries of the city that converge on the Cathedral to the east. The last named of these streets, with its outside stairs, quaint-roofed houses, and College church and tower, has the indescribable

West Port, St Andrews

mediaeval glamour that so fascinates one in many of the Fife towns.

The other noteworthy feature of certain Fifeshire burghs is the old town cross. Quaint specimens may be seen in Kilrenny, Anstruther, Leven, Culross, Kincardine, and many another town. Originally such crosses would of course have a religious significance—would be a reminder of higher things in the midst of daily life. But as time went on—probably because in bargain-making a compact made in such sacred precincts would have a specially binding force—the cross came to be associated with the market and to be the centre of the trade of the little town. The Market Cross of Inverkeithing, a hand-

The Cross, Kincardine-on-Forth

some pillar surmounted by a sun-dial and a unicorn, is considered one of the finest of its kind. The Cross of Cupar was rescued from oblivion some time ago and erected on Tarvit Hill.

The new Corporation Buildings of Dunfermline exhibit a handsome combination of the Scottish Baronial and French Gothic styles; and the decorations of its chambers are worthy of the exterior. Early English architecture is exemplified in St Margaret's Hall. The County Buildings are easily distinguished by their elegant and lofty spire.

The St Andrews Town Hall, in the Scottish style, contains the council room, a police station, and a public hall. The monument erected in 1842 to the memory of the Martyrs who suffered death in Covenanting times, is a well-proportioned obelisk, 45 feet in height. The handsome buildings of the United College that flank a spacious quadrangle form a pleasing contrast to the old church and the tower under which is the main entrance. St Mary's College, a much older building than the United College, with its fine old quadrangle, breathes the air of quiet, almost monastic, seclusion.

24. Architecture—(*d*) Mansions and Famous Seats.

When Defoe visited Fife he remarked that the Scottish kings had more palaces than the kings of England. Two of these are in Fife and they are of unusual interest. Dunfermline, long the Scottish capital and the seat of royalty, has in its palace a noble ruin. It is one of a group of buildings, including Malcolm's Tower, the monastery, and the abbey, that would make any town famous. Far older than the palace is the tower that once crowned the hill hard by, to which Malcolm Canmore brought Margaret as his bride. Its fragmentary remains mark the birthplace of the British Royal House,

United College, St Andrews

for here was born that Princess Matilda whose marriage with King Henry I united the Saxon and the Norman line. The palace, whose stately ruins rise high above the winding stream of Pittencrieff Glen, is probably as ancient as the reign of Alexander III. Edward I is said to have resided here, and on leaving to have ordered the building to be burned. James V is credited with having erected the greater part of the structure the remains of which are yet with us. After the royal family removed to England, the palace was allowed to fall into disrepair. The mullioned windows attest its ancient beauty.

Falkland is inferior only to Dunfermline in its associations with our sovereigns. It was part of the property of the Earls of Fife as early as the twelfth century, but the oldest ruins of which traces yet remain belong to the thirteenth. The castle of Falkland, which has entirely disappeared, must not be confused with the palace. The latter was begun by James III or James IV and completed in 1537 by James V. In life and in death, the last of these kings had a close connection with Falkland. The palace was the favourite residence of James VI. It was the headquarters of the Cromwellian troops in the district. Architecturally the palace may be described as a Renaissance edifice, though it has features distinctive also of the Gothic and the Baronial styles.

In its domestic architecture Fife has been more conservative of the antique parts of its great mansions than many other districts: there have been fewer sweeping changes than in, say, the west of Scotland. Hence if there is less uniformity of design, there is more feeling of

Dunfermline Abbey Gatehouse and Refectory. Ruins of Palace adjoin on the right

union with the past, more of the mellowing influence of time; and in the finest buildings the good taste of owner and architect has shown itself in a careful blending of the new with the old.

Broomhall, the seat of the Earl of Elgin, is a handsome edifice, beautifully situated near the Forth at Limekilns. As the nearest direct male descendant of Robert the Bruce, the Earl has appropriately stored amongst his relics of olden days the sword and helmet of the warrior king. In the case of Fordel, mansion and castle are separate buildings, the older edifice having been for some time in disrepair, though it is now renovated. Both stand amidst picturesque grounds. Close by the Forth in the same district is Donibristle House, which has had the misfortune to be partially destroyed by fire no fewer than three times. The mansion is famous for a finely wrought iron gateway of Flemish workmanship and design.

Originally built in 1694, Raith House, in the Kirkcaldy district, is one of the noted mansions of Fife. Two wings and a fine Ionic portico are modern additions.

Leslie House, near the town of the same name, was built in the time of Charles II. As originally constructed it was quadrangular in form, but a fire destroyed all but a fourth part of it, which now forms the mansion. It contains an interesting portrait gallery. Balbirnie, in the same neighbourhood, is Grecian in style with a façade supported by Ionic pillars at the entrance. Strathendry is an elegant example of the Tudor style of architecture. In the adjoining parish of Kinglassie the principal mansion is Inchdairnie, one of the finest baronial piles in the county.

Wemyss Castle has a remarkably picturesque situation on a lofty cliff overhanging the Forth. Parts of the building are as old as the thirteenth century, but most of it is of recent date and of baronial style. On the opposite side of Largo Bay is Elie House, a large mansion in the Renaissance style with beautiful grounds. Behind Elie and to the north and south respectively of

Leslie House

Colinsburgh are situated Balcarres House and Kilconquhar Castle. The former, a Scoto-Flemish-Gothic structure with fine terraced gardens, is memorable as the home of Lady Anne Lindsay, the gifted authoress of *Auld Robin Gray*. Kilconquhar Castle deserves special mention as one of the most successful instances of the blending of the old and the new in its architecture. The old tower dominates but harmonizes well with the modern additions. Near Pittenweem Kellie Castle stands conspicuously and picturesquely on Kellie Law. Belonging in part to the sixteenth and in part to the early seventeenth century, it has had no external alterations since 1606.

The old fortalice of Balcaskie was transformed towards the end of the seventeenth century into the Jacobean

Wemyss Castle

mansion that is one of the chief architectural ornaments of Fife. Its ceilings are thought even to rival those of Holyrood, and, having the advantage of age, its fine terraces are in exquisite keeping with the manor house. In the grounds are statues of nymphs and fauns appearing at unexpected turns on paths that wind amid bosky groves, and along the principal terrace are ranged busts of Roman emperors—both indications of the Renaissance taste that is displayed in the style of the house.

Melville House, erected by the first Earl of Melville in 1692, has the architectural characteristics of the period. The house has a fine oak staircase and contains some excellent historical portraits. The beech avenue which

Kellie Castle, Pittenweem

forms the chief approach is held to be one of the finest in existence and the grounds are adorned with magnificent old trees. Not far off, in the Parish of Ceres, is the ruin of what must have been one of the most beautiful houses in Fife. Craighall (built in 1637) stands on the summit of a hill of some elevation, and displays in the mere shell of a building that still remains a combination of French

Renaissance work and the Scottish baronial style of architecture.

In the midst of typical Fife scenery lies ancient Earlshall, one of the most picturesque and romantic manor houses of the county. Founded by Alexander Bruce about the middle of the sixteenth century, it is

Craighall Castle, Cupar

more closely to be associated with Sir William Bruce and his wife Dame Agnes Lindsay, who did much to adorn its celebrated hall or drawing-room. Their descendant, Bruce of Earlshall, so notorious in the Killing Time, was another of its proprietors. After lying for many years in ruins it was restored in a style befitting its quaint architecture.

Far different from any of these houses is Crawford Priory, three miles south-west of Cupar. It combines in its stately pile the characteristics of domestic and ecclesiastical Gothic architecture: its appearance gives the impression at once of a castle and a monastery. Originally a castellated edifice, it has a tower with a spire 115 feet high. A newer part of the building, the private chapel, is regarded as one of the most artistic in Scotland. There is no more imposing mansion in Fifeshire. Another fine example of the Gothic style in domestic architecture is Kilmaron Castle, also in the Cupar district. Beautifully situated in a well-wooded park, it forms a striking object in the surrounding landscape.

Falkland House occupies a commanding and beautiful position near the foot of the Lomonds. It is an entirely modern building, having been erected between 1839 and 1844, and ranks amongst the finest residences in the county. Inchrye Abbey is an ornate and spacious mansion situated near the loch of Lindores. Erected by George Ramsay of Inchrye in 1827, it is a fine specimen of Tudor architecture. Want of space alone forbids more detailed notice of such north of Fife houses as Ayton, Mountquhanie, Birkhill, Naughton, St Fort, and Scotscraig.

In the days when it was too far a cry to spend part of the year in London or even Edinburgh, many of the nobility and gentry of Fife had town residences in the Fife burghs. One of these is the so-called "Palace" in Culross, built by Sir George Bruce between 1597 and 1611. Like many another old dwelling, it was conspicuous by standing at right angles to the street. Unimportant though antique-looking externally, it is of peculiar interest in its richly ornamented hall and in the mural decorations of its rooms. James VI is said to have been entertained here on his visit to Scotland in 1617. Still older is Queen Mary's House, St Andrews, which, erected

Old Palace, Culross

in 1523, was occupied for some time by Queen Mary, and afterwards by Charles II in 1650. Abbot Pitcairn's House, Dunfermline, a quaint Z-shaped structure, dating from the thirteenth century and marked with shot-holes, bears the wise motto—

> "sen vord is thrall and thocht is fre,
> keip veill thy tonge, I coinsel the."

25. Communications—Past and Present: Roads and Railways.

The earliest lines of communication must have followed the natural through routes of the district. What are these natural through routes? Tracks running east and west must have followed the valleys of the main streams, as being lines of least resistance; those going north and south, the openings or natural depressions, in the Ochils, Lomonds and other ranges. Drove roads would often cross heights. Besides these there must always have existed a road as nearly as possible parallel, and close, to the shore. When we come to speak of the Fifeshire railways, it will be found that these old natural routes have been pretty closely adhered to.

Another circumstance that must be taken into consideration is the need of constant communication between various important centres, which would lead from the earliest times to the formation of beaten tracks and afterwards of roads. Now Culross, Dunfermline, Kirkcaldy, Markinch, Loch Leven, Falkland, Cupar, St Andrews, and Newburgh—to mention only the most outstanding names—figure prominently in the history of Fife; and from one to another of these places there would from the earliest times be frequent comings and goings. Moreover on the outskirts of the district lie a few important centres—Edinburgh, Stirling, Perth, and Dundee, to reach which from Fife would early involve the construction of bridges or the use of ferries.

We may take it, then, that the older routes through the county are substantially those with which we are now familiar; though, of course, scientific road-making is a modern art.

Let us first take the routes from west to east. Perth connects with Newburgh, Lindores, Cupar, and St Andrews; Stirling with Kinross, and thence through the Howe of Fife with Cupar; and from Stirling another, the shore route, leads by Culross, Dunfermline, Inverkeithing, Burntisland, Kirkcaldy, Leven and Anstruther to Crail and St Andrews. Not less important would be the routes from south to north. Perth would be reached from Dunfermline or Inverkeithing by Kinross and the Wicks of Baiglie, a route over which Scott is so enthusiastic in *The Fair Maid of Perth*, though he deplores, on scenic grounds, the abandonment in favour of the easier Glenfarg road of the older track that climbed the Ochils and commanded the magnificent view of Strathtay, which he describes so eloquently. The kings who lived at Dunfermline might reach their favourite hunting ground at Falkland either by the route that skirts Loch Leven on the west and joins the main road through the Howe of Fife, or the more direct way that passes through Leslie and Markinch and crosses the saddle between that town and Kettle. The same route is joined at Markinch by that other which goes direct from Kirkcaldy to Cupar, Kilmany or Leuchars, and the ferries on the Tay. Edinburgh would communicate regularly with Fife either by the Queensferries or Pettycur, and Dundee by Balmerino, Woodhaven, Seamylnes (Newport) or Ferry-port-on-

Craig. On old milestones was marked the distance from Pettycur as the leading point of approach from East Fife to Edinburgh. There would also be frequent communication by boat between the coast towns on the outer Forth and the opposite shores of the Lothians.

The inconvenience and delay of coaching days are well illustrated in Scott's *Antiquary*, when the rickety "Hawes Fly" broke down and thereby caused Jonathan Oldbuck of Monkbarns and his companion to miss the tide at Queensferry. They had in consequence to spend the night there; and, to make up for lost time, travelled through Fife on the following day by special post-chaise at the rate of eighteenpence a stage. A day and a half was occupied in their journey from Queensferry to Fairport (Arbroath)—a distance now overtaken daily in about a couple of hours. Lovel's well-furnished trunk, be it observed, arrived by sea. That was in the end of the eighteenth century, but even so late as 1840 the Rev. David Anderson of Ceres tells how after crossing the Firth from Newhaven to Largo he found on landing that there was no coach to Cupar and that he had therefore to walk to Ceres carrying his carpet bag!

Dr John Thomson of Markinch, in his account of the agriculture of Fife, a work written near the close of the eighteenth century, deplores the bad state of the Fife roads as one of the chief obstacles to the improvement of the county. But about that time the authorities were beginning to take the matter up: the county was divided into four road districts, and what with good management and the use of the excellent material so easily available in local quarries, the roads of Fife are now amongst the best in the country, and few districts are more liberally provided with this means of communication. Road money was long raised by tolls; but after much vigorous agitation an act was passed in 1878 by which tolls ceased throughout Scotland, the roads being now maintained by a general assessment.

Fife is now well served with railways, all of which form part of the North British system. Two main lines run northwards from Edinburgh by the Forth Bridge, which has almost entirely superseded the Burntisland ferry: one by Dunfermline, Kinross, and Glenfarg to Perth; the other by Kirkcaldy, Cupar, Wormit and the Tay Bridge to Dundee and Aberdeen. The latter forms an integral part of the East Coast route to London. From Leuchars a coast line traverses the East Neuk by St Andrews, Crail, Anstruther, and Leven, and rejoins the main line at Thornton. A small section of the old trunk line goes from Leuchars to Tayport and connects with the Tay Bridge *via* Newport. The most recent addition to the railway system of Fife, the North of Fife line, joins Leuchars with Newburgh and Perth by way of St Fort. Newburgh and Perth may also be reached by rail from Ladybank, while an important branch runs westwards to Stirling through the Howe of Fife. Leslie is reached by a branch from Markinch. Next to the main lines the most important in the county is that between Dunfermline and Thornton with a westward continuation to Alloa, Stirling, and Glasgow, and an eastward to Buckhaven and Leven. It is by means of

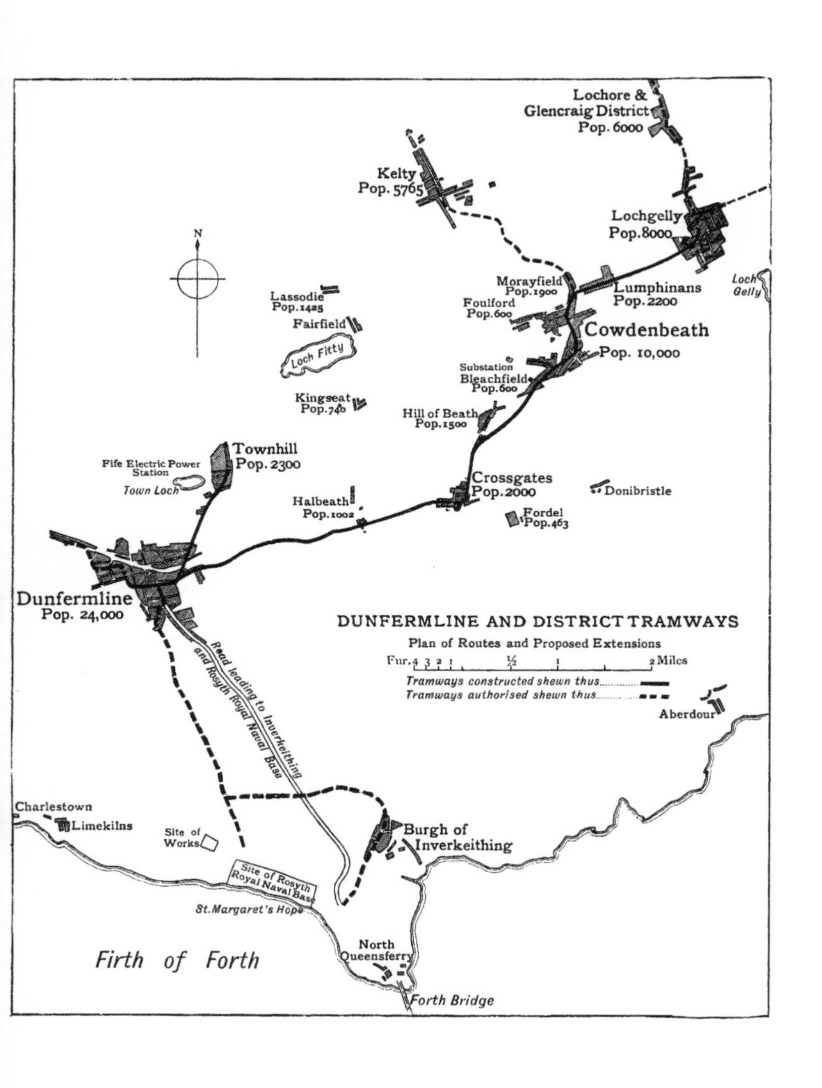
Lochore &
Glencraig District
Pop. 6000
Kelty
Pop. 5765
Lochgelly
Pop. 8000
N
Morayfield
Pop. 1900
Lumphinans
Pop. 2200
Loch Gelly
Lassodie
Pop. 1425
Foulford
Pop. 600
Fairfield
Cowdenbeath
Pop. 10,000
Loch Fitty
Substation
Bleachfield
Pop. 600
Kingseat
Pop. 740
Hill of Beath
Pop. 1500
Townhill
Pop. 2300
Fife Electric Power Station
Town Loch
Crossgates
Pop. 2000
Donibristle
Halbeath
Pop. 1002
Fordel
Pop. 463
Dunfermline
Pop. 24,000
DUNFERMLINE AND DISTRICT TRAMWAYS
Plan of Routes and Proposed Extensions
Fur. 4 3 2 1 ½ 1 2 Miles
Tramways constructed shewn thus
Tramways authorised shewn thus
Road leading to Inverkeithing and Rosyth Royal Naval Base
Aberdour
Charlestown
Limekilns
Site of Works
Burgh of Inverkeithing
Site of Rosyth Royal Naval Base
St. Margaret's Hope
North Queensferry
Firth of Forth
Forth Bridge

this line that coals are conveyed from the rich and populous Dunfermline and Cowdenbeath district to the important docks at Methil and Burntisland.

In addition to its railways, Fife has now two great systems of public tramways, one connected with Kirkcaldy and the other with Dunfermline. The former, which is the older of the two, was extended in 1905 on the formation of the Wemyss and District Tramways Company, Limited. The main line laid down by this company joins Leven, Methil, Buckhaven, Wemyss, Coaltown to Dysart and Kirkcaldy, and taps a district rapidly growing in population and industrial importance.

The tramways of the Dunfermline district are even more extensive. As at present constructed they reach from Dunfermline to Lochgelly, a distance of nine and a half miles; and extensions to Kelty, Lochore, Rosyth, etc., will soon follow. The district, which is the most densely peopled in Fife, has at present a population of more than 70,000. The route already opened serves Dunfermline, Halbeath, Crossgates, Hill of Beath, Bleachfield, Cowdenbeath, Lumphinnans, Morayfield, and Lochgelly. The promoters are the Fife Tramway Light and Power Company, Limited. This company has powers for the supply of electricity over the whole of Fife, excepting the burgh of Kirkcaldy. It is highly probable that these two great systems will be united at no distant date; and a tramway service will then be in operation on a scale unequalled in the east of Scotland.

26. Administration and Divisions—Ancient and Modern.

The administration of Fife may be considered under two heads—local and parliamentary; and it will be convenient to treat the subject in this order.

In the case of an English county it is possible to trace the administration from early Saxon times. In Scotland early local government is more obscure; the country was very unsettled and centuries elapsed before the king had complete control of the separate shires. Counties must at first have been given in charge to warlike nobles able to maintain order and levy contributions in the various districts for the general purposes of the sovereign. These nobles were at first called maormors, and we may regard them as the equivalent of the English earls.

Saxon and Norman customs and forms of government gradually superseded the earlier Celtic ones after the reign of Malcolm Canmore, and so we find that the first Earl of Fife was Ethelred, a son of that monarch. The term Thane of Fife, with which we are familiar in *Macbeth*, was probably a mere translation of Maormor of Fife, the former term being of Saxon origin.

Under the Earls of Fife were local barons called sheriffs-depute who administered the affairs of the county. As time went on an appeal from the decision of these petty governors could be made to courts, roughly equivalent to our circuit courts, presided over by justices-in-ayre.

The earldom of Fife in this older sense was abolished in 1425, and the king conferred the supervision of the county on a resident nobleman of approved worth, in whose family the office became hereditary. It proved a position not only of great importance but also of ample emolument, and for three hundred years it was held by the Earls of Rothes. In 1747 this heritable office, then worth £10,000 a year, was restored to the Crown; and it was enacted that in future sheriffs-depute should be advocates appointed by the sovereign.

As the population of the county increased and the sheriff's duties multiplied, under-sheriffs were appointed, and Fife was divided into three districts: a sheriff-substitute holding court at Cupar, and another residing in Dunfermline with courts there and at Kirkcaldy. The sheriff-depute of 1747 is now called sheriff-principal. Moreover the civil and the military duties of the quondam sheriff are now disjoined. Under the Territorial and Reserve Forces Act, 1907, the Lord Lieutenant is *ex officio* president of the Territorial Association.

Since the Local Government (Scotland) Act of 1889, the administrative and financial affairs of the county, outside royal, parliamentary, and police burghs, are under the charge of a County Council. The councillors, 60 in number, chosen by the electors of the various divisions of the county, hold office for three years. There are 43 electoral divisions (including 8 police burghs), and each returns one member to the Council, the remaining 17 being the representatives of 16 royal burghs including St Andrews, which returns two councillors.

Important powers and duties are conferred on the Council. They have complete charge of the roads, and share with an older body, the Commissioners of Supply, the management of the police force. They administer the Public Health and other important acts, and have had transferred to them certain administrative functions formerly exercised by the Justices of the Peace; for example, the execution of the acts relating to Weights and Measures, Wild Birds, etc. Fife is divided into four districts for purposes of the administration of the Public Health Acts and the maintenance of roadways: these are Cupar, St Andrews, Kirkcaldy, and Dunfermline, which are managed by committees of the County Council together with representatives of the various Parish Councils within each district. The county comprehends 61 quoad civilia parishes and the extra-parochial district of the Isle of May. There are also 16 quoad sacra parishes.

Special committees of the County Council take charge of such matters as Technical Education, Finance, Valuation, and certain harbour, ferry, and fishing interests. The Council also elect 15 of the 21 members of the Fife and Kinross Lunacy Board. While the County Council has charge of those matters which are common to the whole county, important local duties are discharged by the Parish Councils, bodies which are established by the Local Government Act, 1894, to take the place of the old Parochial Boards. The Parish Council has the entire administration of the Poor Law in the Parish, and besides rating for that purpose, it also raises the necessary funds for education, the spending of which, however, is in the hands of the School Boards.

Public elementary and secondary education is mainly directed, under the Scottish Education Department, by popularly elected School Boards, 69 in number.

In the first parliament after the Union in 1707, Fife had three representatives—one for the county; one for a group of burghs consisting of Dysart, Kirkcaldy, Burntisland, and Kinghorn; and one for a second group—Anstruther-Easter, Anstruther-Wester, Crail, Kilrenny, and Pittenweem. St Andrews and Cupar were represented along with Perth, Dundee, and Forfar; and Inverkeithing, Dunfermline, and Queensferry along with Culross and Stirling. After 1832 St Andrews was associated with the Anstruther group. In 1868 the universities of St Andrews and Edinburgh united in sending one member to parliament. In 1885 the county was divided for representative purposes into East and West Fife; the Kirkcaldy Burghs remained as before; the Anstruther Burghs were now known as the St Andrews Burghs and included both that city and Cupar; and the old grouping of the Stirling Burghs was retained; likewise the two universities remained associated as before.

27. The Roll of Honour.

The men and women of Fife whose names figure in history belong to many and varied departments of life: there are kings and queens, princes and statesmen, saints and martyrs, scholars and soldiers, poets and artists.

If not exclusively the royal county of Scotland, Fife was for so many centuries the home of royalty that many of our kings and queens, from Malcolm Canmore and Margaret to James VI and Anne of Denmark, were

Charles I

closely identified with it. They held court at Dunfermline, hunted at Falkland, founded many of its ecclesiastical institutions, granted royal charters to many of its towns, and patronised its sons gifted in arts and letters.

Centuries before James VI ruled over both countries, Edith Matilda, a princess born at Dunfermline, had by her marriage with Henry I united the Saxon and the Norman line; and on the eve of the removal of the court to England there was born in the same city the prince who was afterwards to be the unfortunate Charles I.

Many localities in Fife have been specially identified with Scottish princes. Queen Margaret had close association with Queensferry, St Margaret's Hope, and Dunfermline. Ermengarda, William the Lion's saintly queen, was the founder of Balmerino Abbey. Alexander III, "the tamer of the ravens," met a tragic death at Kinghorn. William Wallace, the Guardian of Scotland, gained a victory over the English near Newburgh. No royal name, except Queen Margaret's, is so intimately associated with Dunfermline as that of Robert the Bruce. In Falkland Palace died the Duke of Rothesay, heir-apparent to Robert III, perhaps starved to death by his "false uncle," the Duke of Albany. All the Jameses found in Falkland Palace a favourite pleasure resort. James V was married to Mary of Guise by Cardinal David Beaton in St Andrews Cathedral. Many of the fortunes and misfortunes of Mary Queen of Scots are associated with places in Fife—Falkland, Cupar, Burntisland, Wemyss, St Andrews, and Loch Leven. Charles I and Charles II both visited it; and the sojourn of the latter was curtailed by the advent of Cromwell.

Several of the prominent statesmen connected with Fife were great ecclesiastics as well. Wardlaw, scion of an old county family, was appointed to the see of

Mary Queen of Scots

St Andrews and is remembered as the founder of the first university in Scotland. Kennedy, his successor, continued the good work and established the College of St Salvator, but is even better known as one of the

Cardinal Beaton

regents during the minority of James III, a time when his authority was unsurpassed in Scotland. Scarcely less influential were the Bethunes or Beatons, James the Archbishop, and David the Cardinal : both did much as statesmen to foster French as against English alliance, and as prelates vigorously to suppress Reformation principles. The murder of Cardinal Beaton is one of the tragedies of Scottish history; and another is that of James Sharp, who more than a century later was as zealous in the Episcopal as Beaton had been in the Catholic cause. St Andrews, or its immediate environs, was the scene of these cruel deeds. Knox, though not born in the county, began his great work as a reformer in St Andrews. Andrew Melville, a native of Forfarshire, became his successor and did work of importance second only to that of Knox in the same district. Maitland of Lethington, craftiest of politicians, the secretary to Mary Queen of Scots, belonged to Dysart; though not a churchman, his name must be mentioned here amongst prominent Fife statesmen. Henderson, minister of Leuchars, took the most conspicuous part in connection with the drawing up of the National Covenant and also of the Solemn League and Covenant, and in the negotiations with the English presbyterians of the famous Westminster Assembly.

By an easy and natural transition we pass from clerical statesmen to some of the great divines. The saintly Samuel Rutherford was principal of St Mary's College, St Andrews. Richard Cameron, founder of the Cameronian sect of Covenanters, was a native of Falkland. In the next century the Erskines, Ebenezer and Ralph, were closely associated with Fife and there led the Secession movement. Almost exactly a hundred years later the leader of the Disruption movement, which resulted in the founding of the Free Church, was the great Thomas Chalmers, a native of Anstruther, whose early work in the ministry is indissolubly associated with the little church of Kilmany. For some time also he was Professor of Natural Philosophy in St Andrews. Another distinguished religious leader was Edward

Thomas Chalmers

Irving, the colleague, while yet a teacher in Kirkcaldy, of the yet more famous Thomas Carlyle. John Glas of Auchtermuchty formed the sect known as the Glassites.

Amongst Fife statesmen whose lot was cast in less dangerous if not less troublous times, we may instance two of the Bruces of Broomhall, descendants of Robert the Bruce, and grandfather and father, respectively, of the present Earl of Elgin. The former, when Ambassador-Extraordinary in Turkey, secured for the British Museum

Alexander Leslie, First Earl of Leven (1582–1661)

the famous "Elgin Marbles," which had adorned the Parthenon at Athens; and the latter, after a brilliant political career in other parts of the empire, became Governor-General of India. About the same time (1860) John Campbell, son of a Cupar minister, raised himself by sheer dint of personal ability and perseverance to the high position of Lord Chancellor of England.

Of famous men of action Sir William Kirkaldy of Grange, near Kinghorn, is one with whose name military history rings in the times of Queen Mary and her son. Fife was also the birthplace of the kinsmen, Alexander and David Leslie, renowned soldiers of Gustavus Adolphus in the Thirty Years' War and leaders of Scottish armies in the seventeenth century. Alexander became Earl of Leven and dwelt on his estate at Balgonie; and David, Lord Newark, spent his old age near St Monans.

A maritime county and one closely associated with the formation of the Scottish navy, Fife in early times produced some famous sailors; as Sir Michael of Wemyss, the first Scottish admiral, and Sir Andrew Wood of Largo, so distinguished in naval warfare with England in the time of James IV. Worthy successors to these in the eighteenth century were Sir Samuel Greig and Sir Philip Durham, also of Largo. Greig ultimately gave his services to Russia, designed the fortress of Cronstadt, and attained a high place in the Russian navy. Durham highly distinguished himself at Trafalgar and captured Guadaloupe in 1815. A descendant of Sir Andrew Wood, he succeeded to his estate at Largo. Alexander Selkirk, of humbler origin, was the son of a shoemaker at Lower Largo. He, of course, owes his largely fictitious fame to the genius of Defoe, who may have first heard of his hero when travelling in Fife on public business.

Fifeshire has produced, or has had as residents within

Crusoe Statue, Lower Largo

its boundaries, many men and women who have taken a conspicuous position in literature, science, and art. Michael Scott, of Balwearie near Kirkcaldy, is one of the earliest of these. Traveller and schoolman, philosopher and magician, he had, like many another of our early scholars, to attain his learning abroad.

Andrew of Wyntoun, eldest of Scottish historians, wrote his riming Chronicle in the Priory of St Serf's, Loch Leven, at a time when its precincts were still within Fife. Robert Henryson, one of the most delightful of early Scottish poets, whose perennial grace and humour still charm us in *Robin and Makyn* and *The Town Mouse and the Country Mouse*, is supposed to have spent the greater part of his life in Dunfermline. Sir David Lyndsay of the Mount, near Cupar, satirised the licentious lives of the Romish clergy in his poems and was shielded from their wrath by James V. Lyndsay's *Satire of the Three Estates* was performed at Cupar in 1552. George Buchanan, to whom James VI owed his classical scholarship, is one of the chief ornaments of St Andrews University. He was Principal of St Leonard's College for some time. In earlier life he was associated with John Major, another famous St Andrews professor. Sir John Scott of Scotstarvit, a great student and patron of letters, founded the chair of humanity in the university in the same century. Robert Fergusson, the poet, a St Andrews student in the eighteenth century, powerfully influenced Robert Burns.

Fife has produced not a few excellent songs and ballads. To Lady Elizabeth Halkett has rightly or wrongly been ascribed the ballad of *Hardyknute*. *Gloomy winter's noo awa'* by Charles Gray of Anstruther, *O why left I my hame* by Robert Gilfillan of Dunfermline, *Where Gadie rins* by Dr John Park of St Andrews, and *The Boatie rows* by Henry Sime are specimens of undoubted Fifeshire authorship. *The March of the Cameron Men* is

George Buchanan

by Miss Campbell of Pitlour; and *Kate Dalrymple* finds its scene and its heroine (though fictitious names are used) at Kinaldy near St Andrews. Perhaps the chief literary glory of the county must be claimed for Lady Anne Lindsay's *Auld Robin Gray*, and William Tennant's *Anster Fair*. Recent contributors to literature are Principal Shairp, Sir Noël Paton, Major Whyte-Melville, and Mrs Oliphant.

Sir William Bruce, who completed Holyrood, may fitly represent architecture. In painting Fife has produced, to mention only the two greatest examples, Sir David Wilkie and Sir Noël Paton. Chief ornament

Adam Smith

amongst its philosophers is Adam Smith, whose *Wealth of Nations* founded the science of political economy. In science there fall to be recorded the names of Sir Andrew Balfour, the great botanist of the seventeenth century, and Professor Playfair, the mathematician.

28. THE CHIEF TOWNS AND VILLAGES OF FIFE.

(The figures in brackets after each name give the population in 1901, and those at the end of the sections give references to the text.)

Aberdour (583), a village and parish in the south-west of Fife, is situated on the Forth and commands fine views of Edinburgh and the Pentlands opposite. St Fillan's Church, in which is the grave of the Rev. Robert Blair, chaplain to Charles I, and Aberdour Castle, formerly the mansion of the Earls of Morton, are objects of interest. Nearly opposite lies the Island of Inchcolm with the ruins of a monastery. There is an oyster-bed in Whitesands Bay, which with whelk-picking and fishing gives employment to a section of the villagers. (pp. 40, 52, 119.)

Anstruther (1691) is the chief fishing town of Fife and the headquarters of the fishery district between those of Leith and Montrose. Cellardyke, or Nether Kilrenny, is a continuation of the town eastwards. Anstruther has a brewery, a tannery, and factories for ropes, sails, oil, and oilskins. A Spanish ship, one of the ill-fated Armada, landed at its harbour in 1588. Many of its inhabitants fell at Kilsyth, and the town was plundered by Cromwell's troops in 1651. Its modern harbour was completed in 1877 after eleven years' work at a cost of over £80,000. Dr Thomas Chalmers and Professor William Tennant, the poet, were natives. (pp. 6, 27, 51, 71, 80, 92, 156.)

Auchterderran (7000). The parish of Auchterderran, partly pastoral but largely mining, contains a number of villages rapidly growing into towns around the Glencraig, Cardenden, Dundonald, and Bowhill collieries; but the largest township in the parish is the burgh of Lochgelly of which since 1891 the population has grown from 4133 to nearly 8000. The parish is one of the most important mining centres in the county. Work goes on unceasingly night and day, and there is a daily output of thousands of tons of coal of high quality. (p. 74.)

Auchtermuchty (665), formerly one of the royal burghs, is about ten miles south-west of Cupar, and has linen factories, a bleach field, a large distillery, sawmills, etc. The district is one of the most fertile in the county. (pp. 103, 157.)

Balgonie, two villages near Markinch, on the banks of the Leven. Balgonie Castle belonged to the Earls of Leven. The district is rich in coal-mines which have existed for centuries; and there are flax-mills. (p. 159.)

Ballingry, a hamlet and parish of West Fife. Lumphinnans is an important mining village (2200).

Balmerino (pa. 576), a village and parish in the north of Fife. The village, which is beautifully situated on the Tay five and a half miles south-west of Newport, used to be a good fishing centre. Its abbey, now in ruins, was founded by the Queen of William the Lion. (p. 40.)

Beath (15,812) is a parish containing many villages that are rapidly developing into towns round such collieries as Hill of Beath (1500), Lassodie (1425), Morayfield (1900) and Kelty (5765). Cowdenbeath (10,000) is the largest town. (p. 88.)

Buckhaven, originally a fishing village, was in 1891 formed along with Methil and Innerleven (or Dubbieside) into a police burgh, the population of the united towns being then 6000. Soon after 1901 this had risen to 9428. The progress of the burgh is largely due to the great dock at Methil.

Buckhaven, Methil, and Innerleven have, in addition to their colliery trade, engineering works, and factories for flax-spinning and net-work, while at Kirkland in the immediate neighbourhood are the works of the Scottish Cyanide Co. Ltd. (pp. 83, 85, 96.)

Burntisland (4846), a royal burgh situated on the Forth, almost directly north of Granton with which it communicates by a ferry. The burgh is one of the greatest coaling ports on the east coast of Scotland. It is also a popular sea-bathing resort. Rossend Castle, which figured long as a military stronghold, stands on an eminence at the west end of the town. The Binn (632 ft.) which rises behind is of exceptional geological interest. (pp. 27, 31, 60, 83, 86, 93, 96, 111.)

Cellardyke, see **Kilrenny.**

Ceres, a small village in the parish (1545) of the same name, is a place of considerable antiquity. It was a burgh of barony and its old bridge is said to have been built before the Battle of Bannockburn. Bridgend is its modern part. In the neighbourhood are Dura Den, Craighall Castle, Struthers House and Scotstarvit Tower. Considerable industry in brown linen is carried on. (p. 15.)

Charlestown (632), a seaport village in Dunfermline parish, is closely connected with Limekilns where there are extensive lime works. It is the outlet of the minerals of the district. (pp. 54, 86, 136.)

Colinsburgh, see **Kilconquhar.**

Collessie, a village in the parish of the same name, has a station on the Ladybank and Perth railway. Near it are a megalith, a tumulus, and other relics of antiquity. (pp. 105, 106.)

Cowdenbeath, see **Beath.**

Crail (1115) a royal and parliamentary burgh in the East Neuk is one of the most interesting and picturesque of the towns of Fife. As early as the ninth century it was a place of considerable commerce, and is still noted in the fishing industry. The parish church, with interesting relics of antiquity, still remains. James Sharp, afterwards archbishop, was its minister. It is now a favourite summer resort and one of its many attractions is the golf-links at Balcomie. (pp. 59, 92, 107, 108, 125, 128.)

Crail, from West

Crossgates, see **Dunfermline.**

Culross (380), made a burgh of barony in 1484 and a royal burgh in 1588, has lost its old importance. At one time it had an extensive trade in coal and salt, and its iron "girdles" rendered its hammermen famous. Its Cistercian Abbey (1217) is partly a ruin, but a portion with a well preserved tower forms the parish church. The "Palace" is one of the most interesting of its architectural relics. (pp. 54, 141, 142.)

Cupar (4511) the capital of Fife is a royal, municipal and parliamentary burgh, and the seat of considerable trade. Its charter dates from 1363. Mysteries and moralities were acted in the old days on School Hill; and many interesting buildings are to be seen in the town and its environs. It does a large trade in corn, and other industries are flax-spinning, dyeing, tanning, and malting. The ancient courts of justice were held here; hence the proverb, "He that will to Cupar, maun to Cupar." It has a sheriff-court and is the seat of the County Council. (pp. 35, 72, 100, 131, 141, 149, 161.)

Dairsie (693), or more correctly Dairsiemuir, is a village about three miles north-east of Cupar. Its castle, a ruin on a piece of rising ground, was once the meeting-place of a Scottish parliament (1335). (pp. 106, 126.)

Dunfermline (26,600), the second largest and in some respects the most important town in Fife, is in the south-west of the county. It has been a royal burgh since the beginning of the twelfth century. The residence of a sheriff-substitute, it is the administrative centre of the western division of the county.

"The town of the crooked linn" was for some time the capital of Scotland, and no place in the county, and few in the country, can vie with it in historical interest. Malcolm's Tower, the Abbey, the Palace and its other ancient and modern buildings have already been noticed. The manufacture of fine table-linen gives it a world-wide reputation. Coal was mined in its vicinity prior to 1291, probably the earliest record in Scotland.

Dunfermline is the principal station on the railway between Edinburgh and Perth; and besides it has direct railway communication with Stirling on the one hand, and the main Fife railway at Thornton on the other. It is also connected by rail with Charlestown, in some sense its port. An extensive tramway system, of which the first section has just been opened, links it with the mining towns in the parishes that adjoin it on the east; and the tram line will be carried to the Rosyth naval base on the south. Already a great drainage system, at an estimated cost of £100,000, is being prepared for the district between Dunfermline and Rosyth; and it is intended to lay out a garden city on that intervening space. Like other mining parishes of the south-west

Dunfermline Abbey, from N.E.

of Fife, that of Dunfermline contains many villages that in some cases are rapidly growing into towns fostered by the collieries. Such are Charlestown (632), Limekilns (688), Halbeath (1002), Crossgates (2000), North Queensferry (360), Crossford, Masterton, St Margaret's Hope (376), Patiemuir, Townhill (2300), Kingseat and Wellwood.

Andrew Carnegie, the well-known native of Dunfermline, founded in 1903 a Trust "to give to the toiling masses of

Dunfermline more of sweetness and light." To the trustees were handed over Pittencrieff Park and Glen, and a capital sum sufficient to produce £25,000 yearly. The agencies carrying out the donor's generous ideals include libraries, reading and recreation rooms, baths, gymnasium, College of hygiene, art classes, and fully equipped school of music; while horticulture and artistic handicrafts are carefully fostered. To promote these schemes, there was expended in 1908 no less than £26,642. (pp. 4, 72, 78, 80, 85, 97, 98, 114, 131, 132, 134, 142, 147, 149, 151, 152, 153, 161.)

Dysart (4000) is a royal and parliamentary burgh a little over two miles north-east of Kirkcaldy, with which, though municipally independent, it practically forms one town. Created a royal burgh by James V it had previously been a burgh of barony, and tradition derives its name from "desertum," the solitary cave of St Serf. Near by is Ravenscraig Castle, and a standing stone is said to mark the spot where a battle was fought in 874 with the invading Danes. Antiquity is writ large on its crooked streets and old buildings, beneath some of which were made deep hiding-holes for smuggled goods in days when legitimate prosperity had temporarily deserted it. It used to have a large trade in salt; and fish-curing, malting, nail-making and brewing were extensively carried on. Flax spinning, the weaving of linen and woollen goods, and more particularly its collieries make it important. Its harbour consists of an outer basin, an inner dock, and a patent slip. (pp. 29, 51, 81, 123, 128, 156.)

Earlsferry, once a royal burgh, goes back to the times of Malcolm Canmore and even to those of Macbeth, fleeing from whom Macduff is said to have taken refuge in a cave near Kincraig Point. It is now practically one with Elie and their joint population is about 2000, which, however, is greatly augmented in summer by the large influx of visitors attracted thither by breezy golf-links and fine sea-bathing. (p. 51.)

Elie, a small police burgh of the East Neuk, attractively situated on the Firth of Forth, is practically connected with Earlsferry, though the two places have separate municipal arrangements. It has a fine natural harbour, so deep and sheltered that General Wade is said to have recommended it to the government of his day as a place suitable for a naval station. His name is perhaps enshrined in Wadehaven a little to the east. (pp. 15, 50, 62.)

Falkland (1000), a small town, once the capital of the stewartry of Fife, is situated at the north-eastern base of the East

Falkland Palace

Lomond Hill. It was long a favourite residence of the kings of Scotland. In consequence, Falkland was made a royal burgh in the fifteenth century and retains in its quaint buildings and streets signs of its former importance. The Palace was completed by James V (1537), and it was there he died. Mary Queen of Scots and James VI frequently resided in it. The Palace was restored some years ago by the late Marquess of Bute. (pp. 4, 134, 141, 152, 153.)

Ferry-Port-on-Craig, or **Tayport** (3325), in the extreme north-east of Fife and directly opposite Broughty Ferry in Forfarshire, derived its importance from the ferry, which is said to be the oldest in Scotland. In 1842 this was purchased by the North British Railway, and along with the Burntisland ferry, formed before the great bridges were built an important link in their line of communication between Edinburgh and the north. The chief industries are jute and linen mills, saw-mills, foundries, a bobbin factory, and valuable salmon and mussel fisheries. It has bathing and golfing facilities. (pp. 35, 48, 57, 61, 77, 91, 93.)

Freuchie (827), a quaint village two miles east of Falkland with a power-loom linen factory. Having shared in olden days the prosperity of Falkland, it now shares its decay. (p. 77.)

Gallatown, see **Kirkcaldy.**

Guard Bridge (524), a village four miles from St Andrews at the mouth of the Eden, which is here spanned by an old six-arched narrow bridge built by Bishop Wardlaw in the fifteenth century. It has brickyards and one of the largest paper mills in the country, which employs so many hands that the village has nearly doubled its population since 1881. (p. 48.)

Inverkeithing (1963), a royal, parliamentary, and police burgh, is situated about a mile and three quarters from the northern end of the Forth Bridge on a bay of the same name. It has a fairly commodious harbour, which could easily be improved. Its shipbuilding yard, tan-works, rope-works, and paper-mill render it a fairly busy place. In the parish were born the Russian admiral Sir Samuel Greig, and the great African missionary, Dr Moffat, father-in-law of David Livingstone. (pp. 52, 80, 93, 100, 131.)

Kelty, see **Beath.**

Kennoway (870) is a village and parish of south central Fife. The village stood on the main road from Edinburgh to Cupar, and has a certain air of antiquity about it. Its former prosperity declined after the days of hand-loom weaving. (p. 88.)

Kettle, or **Kingskettle** (700), is a village of central Fife. Its alternative name points to the fact that it at one time belonged to the Crown. Linen manufacture is carried on to a small extent, but was relatively more important in the days of the hand-loom. (p. 103.)

Kilconquhar (350), a village of the East Neuk, is charmingly situated on the shore of Kilconquhar loch, which is famous for its swans. In the parish of the same name there are besides the hamlets of Colinsburgh (352) and Largoward; and amongst its mansions is Balcarres. (p. 137.)

Kilmany is the name of a village and a parish (502) in the north of Fife. Dr Thomas Chalmers, first moderator of the Free Church of Scotland, was minister of Kilmany (1803–1814), and the place is indissolubly associated with his name. In the parish is Rathillet, home of David Hackston, one of the most aggressive of the Covenanters. He was concerned in the murder of Archbishop Sharp. David Balfour of Montquhanie, in the same district, was one of those who planned the assassination of Cardinal Beaton. (p. 157.)

Kilrenny (2549) is a royal burgh, the more important part of which, Cellardyke, or Nether Kilrenny, is virtually a suburb of Anstruther Easter. (p. 109.)

Kincardine-on-Forth (1010). The decline of this the most westerly of Fifeshire towns dates from the introduction of railways, before which its fine ferry (half a mile wide) was the regular line of communication with the south and west. It does a little shipbuilding; but that and its manufactures of ropes, sails, and woollen goods have greatly declined. It ranks as a burgh of barony. (pp. 56, 57, 93, 130.)

Kinghorn (1568), a royal burgh since the days of Alexander III. It is 12 miles north of Edinburgh, with which it used to communicate by the Pettycur ferry, three quarters of a mile to the west. (pp. 52, 86, 159.)

Kilmany Church

Kinglassie, a village in a parish (1478) of the same name, is situated in the Kirkcaldy district of Fife. Its inhabitants are for the most part weavers, and it has a large power-loom factory. Its antiquities include a sculptured standing stone and the site of a Danish fort; and about 1830 a Roman sword, a battle axe, and some iron spear-heads were found in the alluvial deposits of the Leven. (pp. 29, 86, 136.)

Kingsbarns (pa. 652), a village near the coast seven miles south-east of St Andrews. Its royal castle near the shore contained the barns or granaries of the Palace at Falkland. Hence the name Kingsbarns. (pp. 86, 106.)

Kirkcaldy (34,079), the largest and busiest town in Fife, is a seaport and a royal and parliamentary burgh. "The lang toon" contains within its municipal boundaries Invertiel, Linktown, Pathhead, Sinclairtown, and Gallatown. Its High Street is four miles long. It is specially noted for its linoleum and floorcloth, the manufacture of which was introduced in 1847. Created a royal burgh by Charles I in 1644, the traditions of Kirkcaldy go back to St Columba's time. As a head-port it includes as under-ports or creeks all the harbours from Crail to Aberdour, and its railway station is the principal one between Edinburgh and Dundee. The old tower of its parish church is the one building of antiquarian interest; but it contains many fine modern structures, including no fewer than 25 churches. Adam Smith, the founder of the science of political economy, was born in Kirkcaldy. (pp. 27, 51, 72, 77, 78, 80, 94, 120, 136, 147, 149, 151, 157, 161.)

Ladybank (1340), a small police burgh in central Fife, has an important railway junction from which branch lines go to Stirling *via* the Howe, and to Perth *via* Newburgh. It has linen and malting industries. (pp. 77, 80.)

Largo, the name of a parish and two villages—Upper and Lower Largo—in the south-east of Fife, means "sunny, seaward slope," a designation which well describes its situation on the Forth. Largo Law (965 ft.) is a striking eminence. The two villages together with Lundin Mill or Links, Drummochy, and Temple form practically a small town; Upper Largo—the Kirkton—being slightly inland, the others along the shore, on either side of the Kiel burn. The industries—weaving, a flour mill, an oil work, a net factory, etc.—have declined, and even fishing has nothing of its old importance; but, as summer resorts, Upper Largo and more particularly Lundin Links with its fine golf course are remarkably popular and the permanent population (1000) is then largely augmented. In the district are some singularly interesting antiquities—such as the Standing Stones; and two of its natives, Sir Andrew Wood of the *Yellow Carvel*, and Alexander Selkirk (Robinson Crusoe), have immortalised their birthplace. (pp. 50, 106, 107, 159.)

Lassodie, see **Beath.**

Leslie (3587), a burgh in the west of Fife, is prettily situated on a ridge overlooking the valley of the Leven. Its fine green is said to have been the scene of bull-baiting in the days of old; and the "Bull-Stone" to which the animals were fastened remains as a relic of the barbarous pastime. The industries of the burgh, which are varied and prosperous, include spinning, bleaching and paper-making; and are carried on chiefly at Prinlaws, Fettykil and Strathendry. (pp. 86, 136.)

Leuchars (588), a small village in the north-east of Fife, has a fine Norman apse and lies in the midst of excellent farming land near the mouth of the Motray Water. Leuchars is the railway junction for Tayport and St Andrews. (pp. 100, 119, 156.)

Leven (6000), an ancient burgh of barony situated on the Forth at the mouth of the river Leven, was originally a small weaving and fishing village; but its trade in salmon and trout for which its river was once noted, its extensive brick, salt and malt works, and its hand-looms are now of the past. It has mines, foundries, oil-mills, rope-works, saw-mills, corn-mills, and paper-mills. Its spinning mill is the largest in Fife, and its creosote works prepare telegraph poles for the whole country. (pp. 57, 80, 94.)

Lindores, a small village near Newburgh, has an ancient abbey. In the neighbourhood Wallace defeated the Earl of Pembroke in the battle of Black Ironside. (pp. 20, 46, 107, 120, 141.)

Lochgelly, see **Auchterderran.**

Lundin Links, see **Largo.**

Markinch (1500) is a very ancient place and is said to have been the Celtic capital of Fife. The curious terrace-like markings on a hill, which are now almost obscured by trees, have been ascribed to the Romans. The Norman tower of the old church is one of the finest of its kind in Scotland; and the church itself is believed to be the successor of a Culdee building of the sixth century. In the parish of Markinch are the House of Orr, the birthplace of Cardinal Beaton, and Balgonie Castle where the famous Sir Alexander Leslie ("Crookback Leslie"), the Covenanting general, died. Within the parish there are paper-mills at Balbirnie, Auchmuty, and Rothes; at Cameron Bridge a very large distillery; bleach fields, spinning mills, and collieries at Coaltown, Milltown, Balcurire, and other places. Mining is the growing industry. Windygates was the chief posting-station in East Fife in coaching days. (pp. 80, 116, 124.)

Methil, see **Buckhaven.**

Monimail is a village and parish to the west of Cupar. The parish contains The Mount, the residence, if not the birth-place, of Sir David Lyndsay, the famous satirist (1490–1555). On the Mount Hill, stands a conspicuous monument, erected in memory of the Earl of Hopetoun, the Peninsular hero.

Newburgh is an ancient town and an extensive parish on the Firth of Tay in the extreme north-west of Fife. The town is a royal and police burgh and also a small port. It was created a royal burgh in 1456, but arose in connection with the foundation of Lindores Abbey in the twelfth century. Early in the nineteenth century it had a very considerable trade in linen weaving. The chief industries now are malting, quarrying and the timber trade. In 1871 the population was 2777, it is now only 1505. The district has several singularly interesting relics of antiquity, such as Macduff Cross, Mugdrum Cross, and Lindores Abbey. (pp. 25, 46, 77, 88, 98, 103, 109, 127.)

Newport (2869), which forms with Wormit a police burgh, is beautifully situated on the Tay opposite Dundee and stretches eastwards from the Tay Bridge for more than two miles along the shore of the river. Along with Balmerino and Tayport, Woodhaven and Newport (formerly Seamylns) have for centuries been the ferry stations of northern Fife. The burgh, which is virtually a residential suburb of Dundee, has three railway stations, and an excellent ferry-boat service. (pp. 48, 106.)

Oakley, a well-built village in the western parish of Carnock, was for many years (from 1846) the seat of busy iron-works which are now stopped. In the same parish are Carnock village and Cairneyhill, the latter with linen works. Ironstone, limestone and sandstone abound in the district. (p. 28.)

Pittenweem (2000). The name of this East Neuk fishing-town is said to mean "town of the cave." It is one of the numerous royal burghs of Fife, now bereft of much of their former importance, and was created as such by James V in 1542. Its nucleus was doubtless the Priory founded about 1114. At one time there were thirty breweries in the vicinity but its industry now is fishing. (pp. 27, 51, 138.)

Queensferry, North (360). The ferry which for centuries was the chief means of communication between Edinburgh and central Fife, is now almost entirely superseded by the Forth Bridge, which joins the two Queensferries. North Queensferry, in the parish of Dunfermline, has a picturesque situation on the Ferry-hill peninsula. (pp. 8, 52, 88, 153.)

Rosyth, see **Dunfermline** (and pp. 7, 54, 121).

St Andrews (7621), a royal burgh, market and university town and seaport. For ages the ecclesiastical metropolis of Scotland, it is in many respects the most important town in the county. In Pictish times the district in which it is situated appears to have been named Muckross ("boar's promontory") from its being a *cursus apri*, or boar chase. This fact probably remains enshrined in the word Boarhills, the name of a neighbour-

Chained Bible, St Mary's College, St Andrews

ing village. The date of the earliest religious settlement in the district has been placed as early as 347 A.D. St Andrew appears to have been adopted as the national saint of Scotland about the middle of the eighth century. The Culdees must, however, have firmly established themselves in St Andrews in succession to Abernethy before the introduction of Roman christianity, for the rivalry between the two churches in this part of the country was long and stubborn. Ultimately, of course, Rome got the upper hand, and there was a succession of no fewer than thirty-six bishops before Pope Sixtus IV in 1466 appointed Patrick Graham the first archbishop and created St Andrews the metropolitan see of Scotland. Meanwhile the city which came into existence in connection with the Cathedral had been made a royal burgh in 1140 by David I. The bishops and archbishops exercised wide sway in Scotland and ranked next to the royal family. In pre-Reformation times, St Andrews was a place of great trade, but thereafter its prosperity waned considerably. Its university, the earliest in Scotland, was founded in 1411. By 1697 the fortunes of St Andrews had fallen so low that a proposal was made to remove the university to Perth. Nor did prosperity return till Sir Hugh Lyon Playfair set vigorously to work about the middle of last century to restore its former grandeur and secure its amenities. In this he was aided by the opening of the railway in 1853. Since then St Andrews has increased year by year in good fortune until it is now one of the finest and most fashionable seaside resorts in Britain. It owes much to the Royal and Ancient Golf Club founded in 1754 and patronised in 1837 by King William IV. The Cathedral, founded in 1159 and destroyed by the rabble in 1559, was taken in charge by government in 1826 and is now carefully looked after. In 1860 the foundations of the still older church on Kirk Heugh were discovered. Of the Priory (1144), one of the finest monastic institutions in Europe, there remain only the extensive walls and the gateways or Pends. Two other monasteries existed in St Andrews—those of the Dominicans and the Greyfriars: a beautiful ruin of the former in South Street is their only remaining relic. The Castle was half stronghold, half archiepiscopal residence, and has suffered repeatedly in war and from the ravages of the sea. (pp. 4, 15, 28, 32, 35, 48, 57, 58, 59, 72, 97, 98, 101, 109, 116, 126, 130, 132, 142, 151, 155, 156, 161, 162.)

St Davids, a village near Inverkeithing at the terminus of the Fordel mineral railway, has a large export trade in coal.

St Monans, or **Abercrombie** (1898), a small fishing town of the East Neuk, two and three-quarters miles south-west of Anstruther, has a good harbour, partly natural and partly artificial, which is large enough to accommodate three or four trading vessels and 100 fishing boats. (pp. 51, 60, 119.)

Saline, a parish and village of south-western Fife. Apart from agriculture the inhabitants find employment in the abundant coal, lime and iron of the district. (p. 28.)

St Andrews Cathedral, West Front

Strathmiglo, a village and parish in the north-west of Fife in the valley of the Eden. The village, which is a burgh of barony, has linen works and a bleach-field. It possesses a fine green and a handsome church with an octagonal spire 70 feet in height. Cairns and tumuli in the parish seem to bear evidence of some great battle.

Tayport, see **Ferry-Port-on-Craig.**

Thornton (1147) is an important railway junction in central Fife with branch lines to Dunfermline on the west, and to Buckhaven and Anstruther on the east. There are collieries in the neighbourhood; and the place is one of growing importance. (p. 146.)

Torryburn, a parish in the south-west of the county, contains a small sea-port, which was once the sea-port of Dunfermline. (p. 54.)

Boiling Caldron, St Monans

Wemyss (2177). The name of this parish and town ot south Fife means caves, and the appellation is justified by the eight or ten caves that exist in the vicinity. One of them, the Glass Cave, was the seat ot the earliest glasswork in the country, though it is no longer used for the purpose. The towns of East and West Wemyss—a burgh of barony—engage in fishing, brewing, and the manufacture of linen; but the collieries of the parish, which are very extensive, are destined to insure its prosperity. They are carried on at the above villages and also at Coaltown of Wemyss, Earlseat, Muiredge, and other places in the parish. The harbour of West Wemyss is one of the most picturesque in Fife. In Wemyss Castle, Mary Queen of Scots first met Darnley in 1565. (pp. 29, 51, 91, 105, 137.)

Wormit, see **Newport** (and pp. 25, 34, 48).

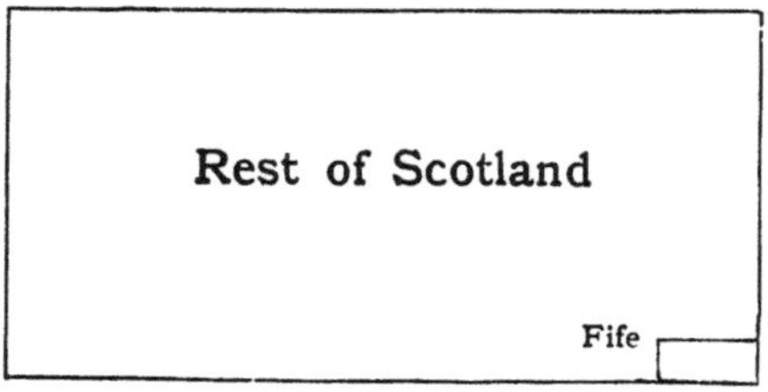

Rest of Scotland

Fife

Fig. 1. Area of Scotland compared with that of Fife

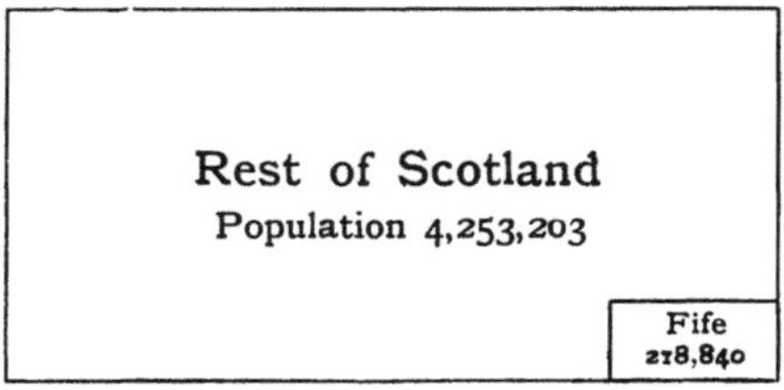

Rest of Scotland
Population 4,253,203

Fife
218,840

Fig. 2. Population of Fife compared with that of Scotland in 1901

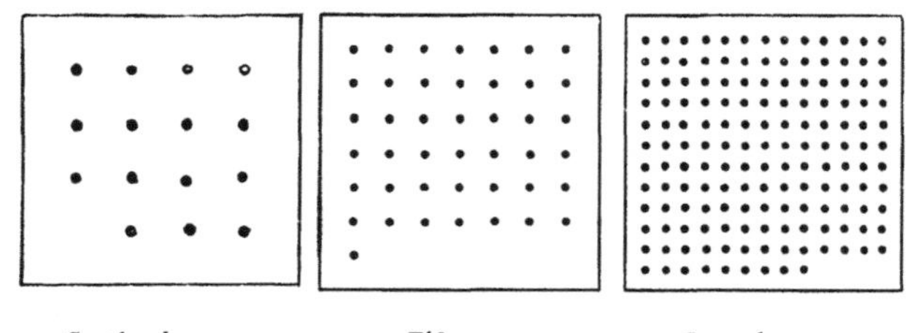

Scotland, 151 Fife, 434 Lanark, 1524

Fig. 3. Density of Population of Fifeshire compared with that of Scotland and with that of Lanarkshire

(*Each dot represents* 10 *persons*)

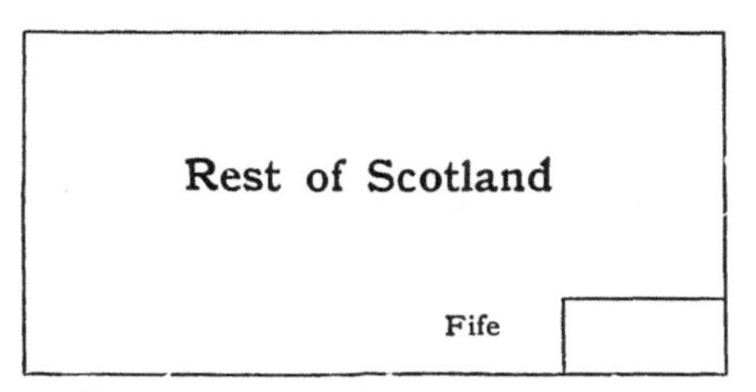

Fig. 4. Comparison of the areas under cultivation in Fife and in Scotland

Fig. 5. Comparison of the areas of Cereals (wheat, barley, oats) grown in Fife and in Scotland

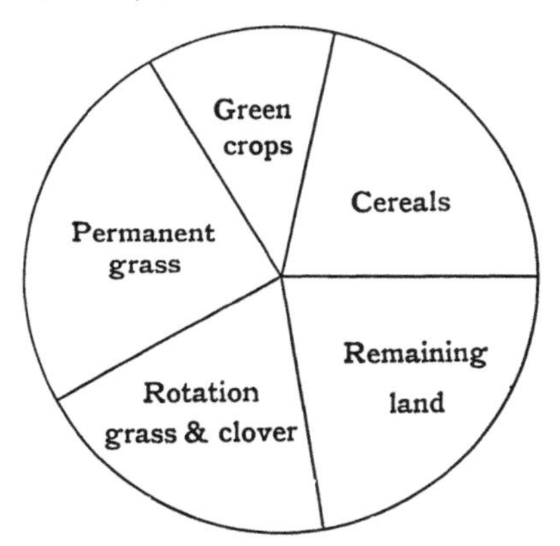

Fig. 6. Proportionate areas of Fife under Cereals, Pastures, etc.

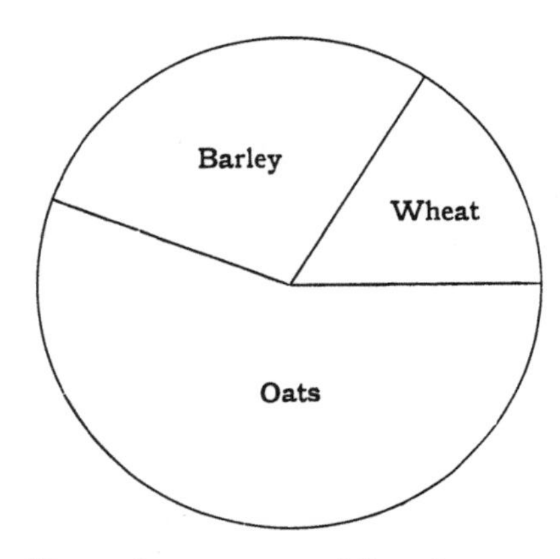

Fig. 7. Proportionate areas of Cereals grown in Fife

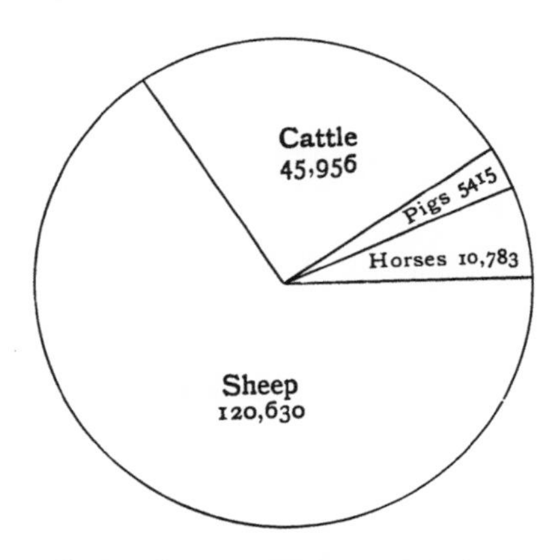

Fig. 8. Proportionate numbers of Animals in Fife

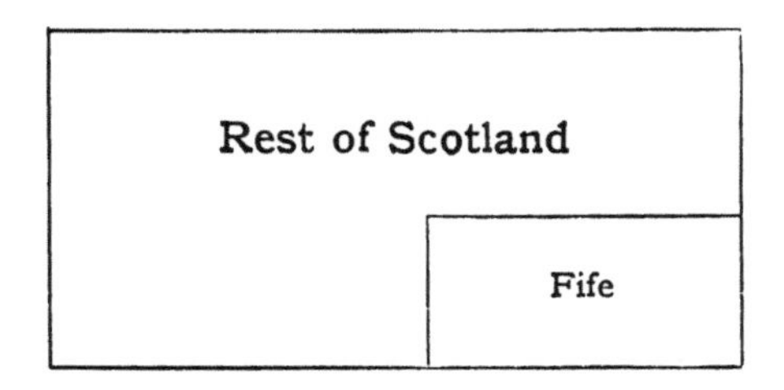

Fig. 9. Comparison of Coal outputs for Fife and Scotland, 1905

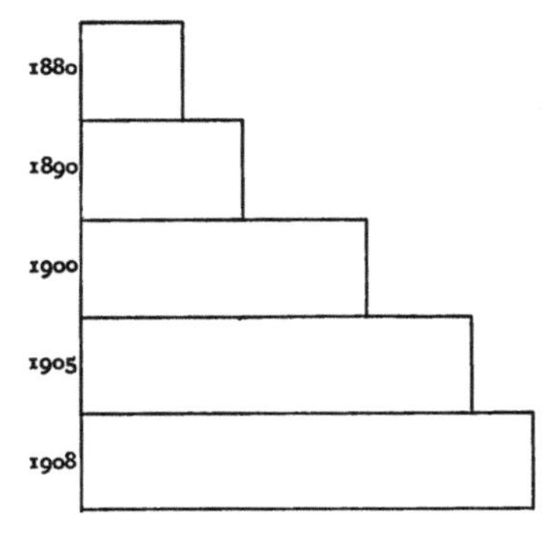

Fig 10. Growth of output of Coal in Fife

www.ingramcontent.com/pod-product-compliance
Ingram Content Group UK Ltd.
Pitfield, Milton Keynes, MK11 3LW, UK
UKHW042143280225
455719UK00001B/67